Conference Proceedings of the Society for Experimental Mechanics Series

Series Editor

Kristin B. Zimmerman, Ph.D.
Society for Experimental Mechanics, Inc.,
Bethel, CT, USA

More information about this series at http://www.springer.com/series/8922

La Vern Starman • Jennifer Hay • Nikhil Karanjgaokar
Editors

Micro and Nanomechanics, Volume 5

Proceedings of the 2016 Annual Conference on Experimental and Applied Mechanics

Springer

Editors
La Vern Starman
Air Force Research Lab
Wright-Patterson AFB, OH, USA

Jennifer Hay
Nanomechanics, Inc.
Oak Ridge, TN, USA

Nikhil Karanjgaokar
Worchester Polytech Institute
Worchester, MA, USA

ISSN 2191-5644 ISSN 2191-5652 (electronic)
Conference Proceedings of the Society for Experimental Mechanics Series
ISBN 978-3-319-42227-5 ISBN 978-3-319-42228-2 (eBook)
DOI 10.1007/978-3-319-42228-2

Library of Congress Control Number: 2016948854

Printed on acid-free paper

This Springer imprint is published by Springer Nature
The registered company is Springer International Publishing AG Switzerland

Preface

Micro and Nanomechanics represents one of ten volumes of technical papers presented at the 2016 SEM Annual Conference and Exposition on Experimental and Applied Mechanics organized by the Society for Experimental Mechanics and held in Orlando, FL, June 6–9, 2016. The complete proceedings also includes volumes on *Dynamic Behavior of Materials; Challenges in Mechanics of Time-Dependent Materials; Advancement of Optical Methods in Experimental Mechanics; Experimental and Applied Mechanics; Mechanics of Biological Systems and Materials; Mechanics of Composite and Multifunctional Materials; Fracture, Fatigue, Failure and Damage Evolution; Residual Stress, Thermomechanics and Infrared Imaging, Hybrid Techniques and Inverse Problems; and Joining Technologies for Composites and Dissimilar Materials.*

Each collection presents early findings from experimental and computational investigations on an important area within experimental mechanics, micro and nanomechanics being one of these areas.

The micro and nanomechanics fields are specialized scientific areas that involve miniaturizing conventional-scale components and systems to take advantage of reduced size and weight and/or enhanced performance or novel functionality. These fields also encompass the application of principles ranging from the micron scale down to individual atoms. Sometimes these principles borrow from conventional-scale laws but often involve new physical and/or chemical phenomena that require new behavioral laws and impart new properties to exploit. Studying how mechanical loads interact with components of these scales is important in developing new applications, as well as assessing their reliability and functionality. Establishing this symposium at the Annual Meeting of the Society for Experimental Mechanics provides a venue where state-of-the-art experimental methods can be leveraged in these endeavors.

The 2016 symposium is the seventeenth in the series and addresses pertinent issues relating to design, analysis, fabrication, testing, optimization, and applications of micro and nanomechanics, especially as these issues relate to experimental mechanics of microscale and nanoscale structures.

It is with deep gratitude that we thank the organizing committee, session chairs, authors and keynote speakers, participants, and SEM staff for making the 17th *International Symposium on Micro and Nanomechanics (ISMAN)* a valuable and unforgettable experience.

Wright-Patterson AFB, OH, USA — La Vern Starman
Oak Ridge, TN, USA — Jennifer Hay
Worchester, MA, USA — Nikhil Karanjgaokar

Contents

Chapter 1
A Stochastic Multi-Scale Model for Predicting MEMS Stiction Failure

T.V. Hoang, L. Wu, S. Paquay, J.-C. Golinval, M. Arnst, and L. Noels

Abstract Adhesion is an important phenomenon in the context of MEMS for which the surface forces become dominant in comparison with the body forces. Because the magnitudes of the adhesive forces strongly depend on the surface interaction distances, which in turn evolve with the roughness of the contacting surfaces, the adhesive forces cannot be determined in a deterministic way. To quantify the uncertainties on the structural stiction behavior of a MEMS, this work proposes a "stochastic multi-scale methodology". The key ingredient of the method is the evaluation of the random meso-scale apparent contact forces, which homogenize the effect of the nano-scale roughness and are integrated into a numerical model of the studied structure as a random contact law. To obtain the probabilistic behavior at the structural MEMS scale, a direct method needs to evaluate explicitly the meso-scale apparent contact forces in a concurrent way with the stochastic multi-scale approach. To reduce the computational cost, a stochastic model is constructed to generate the random meso-scale apparent contact forces. To this end, the apparent contact forces are parameterized by a vector of parameters before applying a polynomial chaos expansion in order to construct a mathematical model representing the probability of the random parameters vector. The problem of micro-beam stiction is then studied in a probabilistic way.

Keywords Stiction • Adhesive contact • Random surface • Multi-scale contact • Uncertainty quantification

1.1 Introduction

Stiction is the common failure in MEMS in which two micro surfaces permanently adhere together due to the adhesive forces such as capillary forces and van der Waal forces. The stiction failure of micro cantilever beams is illustrated in Fig. 1.1a, in which the cantilever beams are stuck on their substrate. In the present work, only the humid stiction failure resulting from the capillary forces is considered.

In MEMS, because of the comparable length of the two scales, the surface roughness (nanometres) and the ranges of the adhesive forces (nanometres), the interaction involves only the highest asperities of the rough surfaces, see Fig. 1.1b [1]. Moreover, due to the scale separation between the ranges of the adhesive forces (nanometres) and the structural displacements (micrometres), the effective contact regions are much smaller than the structural dimensions. For instance, in the case of micro cantilever beams at failure configuration, the effective contact region locates only around the crack tips which is defined as the separating points between the unattached part, for which there is no more interaction forces, and the attached part, see Fig. 1.1a [2]. Therefore, there exist uncertainties in the adhesive behaviors of the micro structures. For instance, in the case of the micro cantilever beam failure, the crack lengths, defined as the length of the unattached part at failure configuration, see Fig. 1.1a, can suffer from a scatter [3–6].

In order to predict the probabilistic behavior of micro structures involving adhesion, the stochastic model-based multi-scale method developed in [2] is applied. This method constructs a multi-scale contact model in which the meso-scale apparent contact forces accounting for the contacting surface topology are integrated as the contact laws into a FE model of the structural scale structure [7]. To predict the probabilistic behavior of the meso-scale apparent contact forces due to surface roughness, contact surfaces are characterized using the power spectral density (PSD) function. Realizations of the contact surfaces are then generated, and the corresponding apparent contact forces are evaluated [8]. Once the distribution of the

T.V. Hoang • L. Wu • J.-C. Golinval • M. Arnst • L. Noels (✉)
Department of Aerospace and Mechanical Engineering, University of Liège, Allée de la Découverte 9, Quartier Polytech 1, B-4000 Liège, Belgium
e-mail: L.Noels@ulg.ac.be

S. Paquay
Open Engineering SA, Rue Bois Saint-Jean 15/1, B-4102, Seraing, Belgium

L. Starman et al. (eds.), *Micro and Nanomechanics*, Conference Proceedings of the Society for Experimental Mechanics Series,
DOI 10.1007/978-3-319-42228-2_1

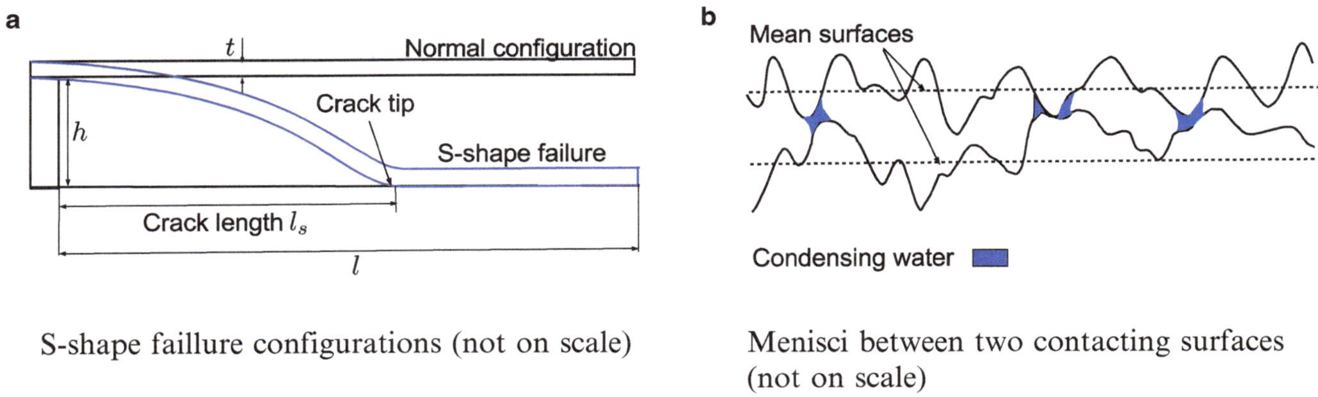

Fig. 1.1 The humid stiction failure of micro cantilever beam: (**a**) the S-shape failure configuration, (**b**) the water condensing between the two contacting rough surfaces

meso-scale apparent contact forces has been obtained, the uncertainties have to be propagated upto the upper scale. The usual method is the direct Monte-Carlo simulation (MCS) which consists in explicitly evaluating for each structural realization a large number of apparent contact forces and in integrating them into the upper scale FE model in order to determine a set of structural behaviors, from which their probabilistic properties are identified. As the evaluation of the apparent contact forces is computationally expensive, the direct MCS method is not efficient. To avoid this disadvantage, the stochastic model-based method constructs a generator for these forces which is built using a relatively small number of explicitly evaluated apparent contact forces. Thanks to the combination of the multi-scale contact model with the apparent contact forces stochastic model, the probabilistic behavior of the adhesion can be determined with an acceptable computational cost.

The structure of this paper is developed as follows. In Sect. 1.2, the physical aspect of capillary forces is recalled. In Sect. 1.3 the stochastic model-based multi-scale method [2] is summarized. In Sect. 1.4, the numerical results are illustrated and compared with the experimental results obtained from [3]. The last section, Sect. 1.5, is the conclusion.

1.2 Humid Stiction of Rough Surfaces

In humid conditions, between the two hydrophilic contacting surfaces, there exists condensing water which results into a negative pressure, named Laplace pressure, and pulls the two contacting surfaces together. The Laplace pressure is evaluated by

$$\Delta P = \frac{\gamma_{\mathrm{LG}}}{r_K} = \frac{\mathcal{R}T \ln RH}{V_m}, \tag{1.1}$$

where γ_{LG} is the liquid vapor energy, V_m is the liquid molar volume, $\mathcal{R}$ is the universal gas constant, and T is the absolute temperature, and RH is the relative humidity. There are two parameters characterizing the geometry of the menisci: the contact angles $\{\alpha_1, \alpha_2\}$, depending on the surfaces properties, and the Kelvin radius r_K given by Cai and Bhushan [9]

$$r_K = \frac{\gamma_{\mathrm{LG}} V_m}{\mathcal{R}T \ln RH} < 0. \tag{1.2}$$

For nano rough surfaces, because the radius of a contacting asperity is much bigger than the Kelvin radius, see Fig. 1.2a, the maximum contact distance at which water can condense between two bodies, defined as the capillary range, is evaluated by

$$h_C = -2r_K(\cos(\alpha_1) + \cos(\alpha_2)). \tag{1.3}$$

Figure 1.2b illustrates the evaluation of the Laplace pressures and of the condensing water heights with humidity levels for contact angles $\alpha_1 = \alpha_2 = 0$.

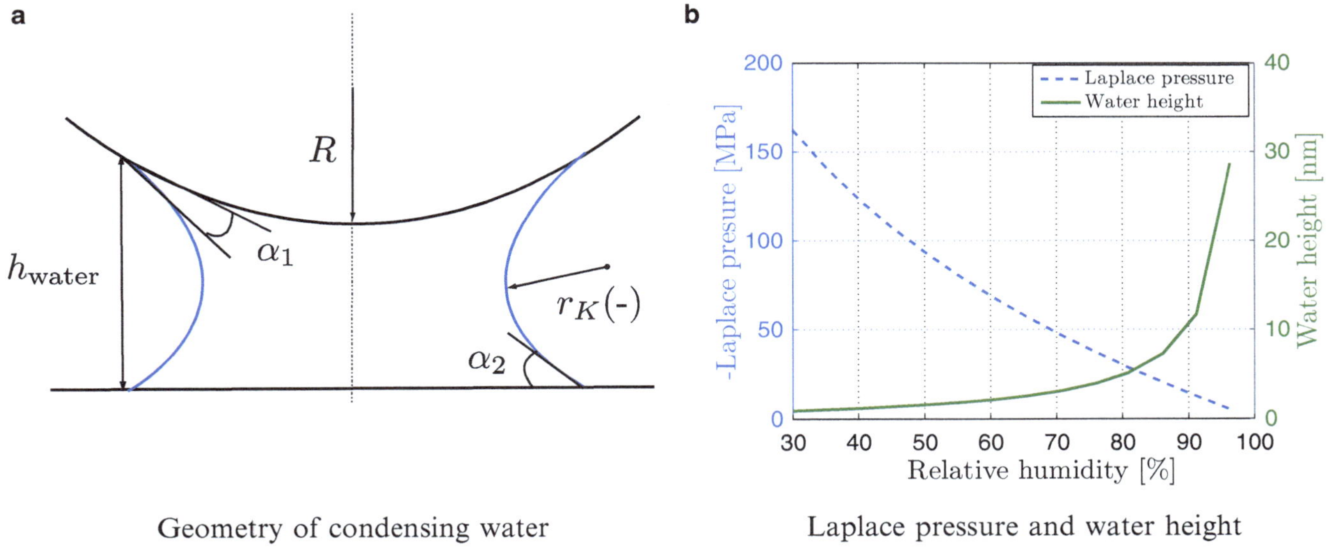

Fig. 1.2 The geometry of a meniscus (**a**), and the evolution of meniscus height and Laplace pressure in terms of humidity levels (**b**)

Fig. 1.3 The meso-scale contact problem

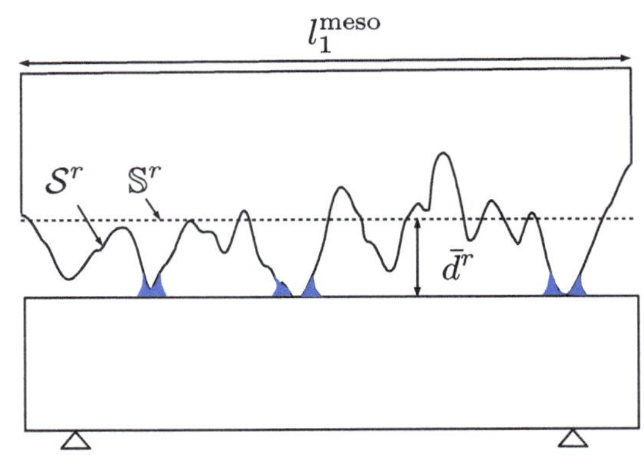

It can be observed that the Laplace pressure is much smaller than the Young's modulus of hard materials such as polysilicon. As a result, at the equilibrium, the physical contact area is much smaller than the apparent area, the typical area ratio being $\mathcal{O}(1\,\%)$. Moreover, the condensing water heights are comparable with the roughness of the typical MEMS surfaces [3, 10]. Therefore, there exist uncertainties on the contact problem between two rough surfaces.

1.3 Modelling Method

In the following the stochastic model-based multi-scale methodology, developed by the authors in [2] to quantify the uncertainty of the humid stiction failure of MEMS structure, is summarized. In this section, the contact problem between two rough surfaces $\mathcal{S}_1, \mathcal{S}_2$ is simplified into the contact between an equivalent surface $\mathcal{S}$ with a plane [11].

In that method, the multi-scale technique is used and consists in replacing the random rough contact surface $\mathcal{S}$ by its mean surface $\mathbb{S}$ and associating it with the contact laws, which are the apparent adhesive contact forces evaluated using the meso-scale contact problem, see Fig. 1.3.

The surfaces $\mathcal{S}$ ($\mathbb{S}$) are partitioned into N_p surfaces $\mathcal{S}^r$ ($\mathbb{S}^r$) ($r \in \{1, \ldots, N_p\}$) with size $l_1^{meso} \times l_2^{meso}$, see Fig. 1.3, for which the variation of the displacement of the mean surface $\mathbb{S}^r$ is negligible in comparison with the roughness of the surface $\mathcal{S}$. Let us define $\bar{d}^r$ the normal distance from the half plane to the mean surface $\mathbb{S}^r$. We consider the meso-scale contact problem in which the displacement of the mean surface $\mathbb{S}$ is prescribed at distance $\bar{d}^r$ from the half plane, see Fig. 1.3. The meso-scale apparent contact force is defined as the average normal contact force applied on the rough surface $\mathcal{S}^r(\theta)$ in the meso-scale contact problem, and is given by

$$\bar{f}^r(\bar{d}^r, \theta) = \frac{1}{A_{\mathbb{S}^r}} \int_{\mathcal{S}^r(\theta)} f(\mathbf{x}, \bar{d}^r, \theta)\mathrm{d}\mathbf{x}, \tag{1.4}$$

where θ is defined on the probability space $(\Omega, \mathfrak{F}, P)$, and $A_{\mathbb{S}^r}$ is the area of surface $\mathbb{S}^r$. Using the meso-scale apparent contact forces to represent the lower scale contact under the form of the contact law, the structural behavior is efficiently determined by conventional methods such as the FE method.

In the following, the probabilistic method [8] used to evaluate the random apparent contact forces Eq. (1.4), and the stochastic model [2] used to represent these forces in the uncertainty quantification process are summarized.

1.3.1 The Random Apparent Contact Forces

In order to evaluate the random apparent contact forces for which the rough surfaces are accounted, the contacting surfaces are characterized using a PSD function from which a surface generator is constructed. Once the realizations of the contacting surfaces, see e.g. Fig. 1.4a, with the size of interest $l_1^{\mathrm{meso}} \times l_2^{\mathrm{meso}}$ are generated, the corresponding apparent contact forces are evaluated using the semi-analytical contact method developed in [8], which consists in 4 steps:

(i) Identifying the physical contacting asperities using the generated contact surface topologies;
(ii) Determining the repulsive contact forces and the physical contacting areas using the Hertz contact model [12] applied on the sphere-fitted profile of the identified asperities, see Fig. 1.4b;
(iii) Evaluating the adhesive contact forces: first, using the physical contacting area and the surface topology, the area on which the water condenses can be identified, then the adhesive forces are calculated from the water condensing area and the Laplace pressure;
(iv) Evaluating the contact forces from the adhesive and repulsive forces.

1.3.2 The Stochastic Model of the Apparent Adhesive Contact Forces

To quantify the uncertainties of the structural behavior due to the involved adhesive contact, there are two possible methods that can be applied: the direct MCS method and the stochastic model-based method [2], illustrated in Fig. 1.5. These two methods have the same objective which is to evaluate a set of N_{MC} structural behaviors corresponding to N_{MC} realizations of contacting surfaces $\mathcal{S}$. The direct MCS method evaluates explicitly the $N_{MC} \times N_p$ apparent contact forces, with N_p the number

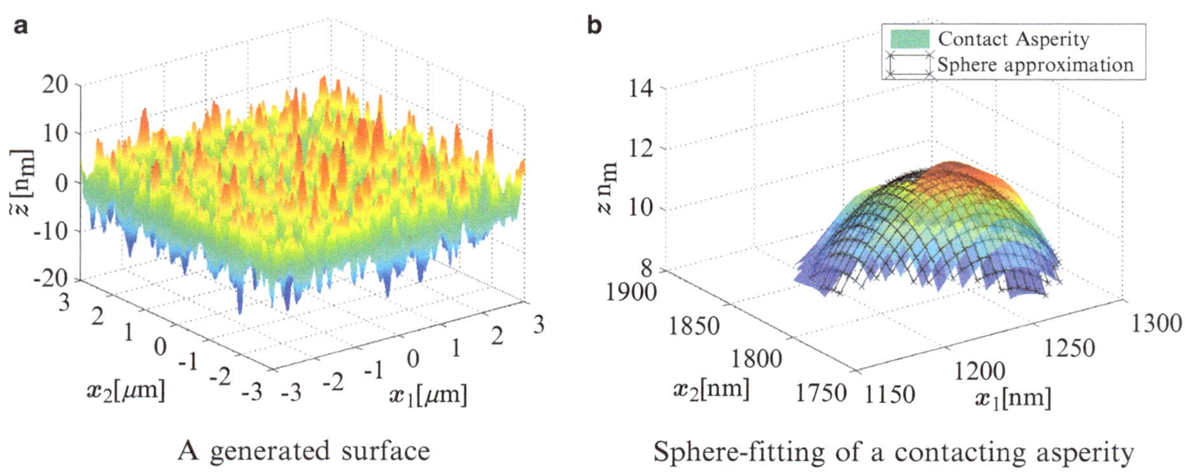

Fig. 1.4 A generated surface using the proposed PSD function in Sect. 1.4 (**a**), and the spherical approximation of a physical contacting asperity (**b**)

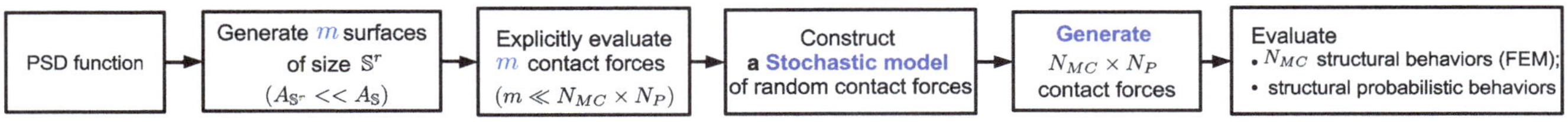

Fig. 1.5 The stochastic model-based multi-scale method used to evaluate the probabilistic behaviors of structures involving adhesive contacts

of contact integration points for each realization while in the stochastic model-based method, a generator for these forces is constructed with the input of m explicitly evaluated apparent contact forces ($m \ll N_{MC} \times N_p$). Because the computational cost required to evaluate the apparent contact forces is hight, the stochastic model-based method is more computational efficient.

In [2], the stochastic model of random apparent contact forces is constructed from two main ingredients:

(i) The representation of the contact forces (distance-force curves) by a Morse potential parametrized by 4-dimensional parameters vectors $\mathbf{v}$, see [2] for details;
(ii) The representation of the random parameters vectors $\mathbf{V}$ by a stochastic model based on the generalized polynomial chaos expansion (gPCE) model, that is,

$$\mathbf{V}^{PC} = \mathcal{T}(\Xi), \tag{1.5}$$

where Ξ was chosen as a 4-dimensional uniform distribution random vector, and the transformation $\mathcal{T}$ constructed from explicit functions, such as polynomials, which are evaluated at a negligible computational cost.

1.4 Numerical Results

The stochastic multi-scale methods are applied to quantify the uncertainties of the humid stiction phenomenon of polysilicon micro cantilever beams, as illustrated in Fig. 1.1.

1.4.1 Problem Setting

The geometry of the cantilever beams is given by the length $l = 1500\,\mu$m, the width $w = 30\,\mu$m, the thickness $t = 2.62\,\mu$m, and the height from substrate $h = 1.9\,\mu$m. Both cantilever beam and its substrate are made of polysilicon for which the Young's modulus $E = 163$ GPa, and water contact angles $\alpha_1 = \alpha_2 \simeq 0^0$. That configuration was experimentally studied and reported in [3].

In this work, we use the PSD function proposed by the authors in [8], constructed using the experimental data reported in [3]. That PSD function is given by (Fig. 1.6)

$$Z(\zeta_r) = \begin{cases} s_Z^0 & \text{if } 0 \leq \zeta_r < \zeta_r^0\ ; \\ s_Z^0 \left(\dfrac{\zeta_r}{\zeta_r^0}\right)^{\log_{10}\left(\frac{s_Z^1}{s_Z^0}\right)/\log_{10}\left(\frac{\zeta\text{max}}{\zeta_r^0}\right)} & \text{if } \zeta_r^0 \leq \zeta_r \leq \zeta\text{max}\ ; \\ 0 & \text{if } \zeta\text{max} < \zeta_r\ ; \end{cases} \tag{1.6}$$

where $\zeta_r = \sqrt{\zeta_{x_1}^2 + \zeta_{x_2}^2}$, $\zeta_{\max} = 289\,\text{rad}\,\mu\text{m}^{-1}$, $\zeta_r^0 = 19.4\,\text{rad}\,\mu\text{m}^{-1}$, $s_Z^0 = 6.46 \times 10^3\,\text{nm}^4$, and $s_Z^1 = 6.29 \times 10^{-3}\,\text{nm}^4$. The surfaces constructed using the proposed PSD function are isotropic, stationary, and have the root mean square roughness of rms $= 3.4$ nm. The contact surfaces are assumed to have a Gaussian distribution and can be generated using the methodology developed in [8]. A realization is illustrated in Fig. 1.4.

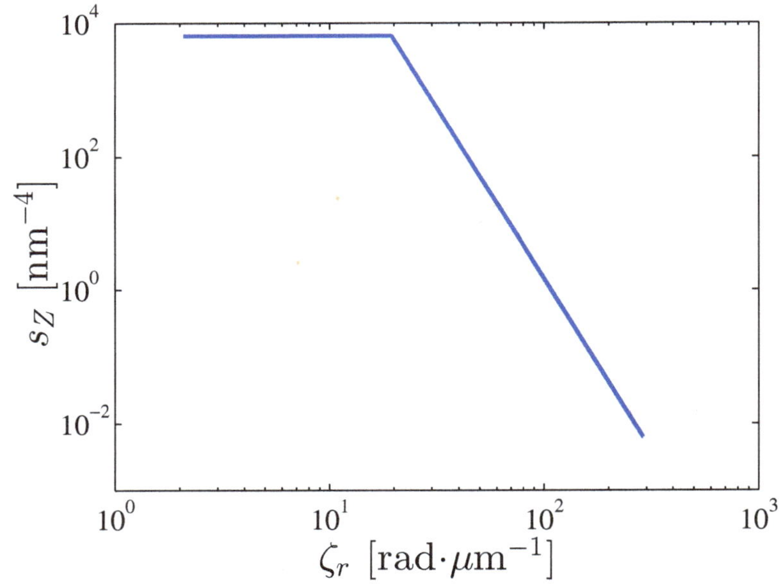

Fig. 1.6 The constructed PSD function corresponding to the experimental data reported in [3]

1.4.2 Implementation

The stochastic model-based multi-scale method is implemented in 3 steps:

(i) Constructing the stochastic model of the apparent contact forces, Eq. (1.5), which consists in generating $m = 5000$ surfaces with size $w \times l^{\text{meso}}$ with $l^{\text{meso}} = 3\,\mu\text{m}$ for which the corresponding apparent contact forces are evaluated using the probabilistic contact model summarized in Sect. 1.3.1, and then in implementing the stochastic model Eq. (1.5) using the method developed in [2];
(ii) Evaluating a set of N_{MC} structural behaviors of the cantilever beam, e.g. crack lengths, which consists in generating $N_{MC} \times N_p$ apparent contact forces using the constructed stochastic model, see Fig. 1.7 which illustrates 10 generated contact forces, and in determining the N_{MC} corresponding structural behaviors by inserting these contact forces as contact laws in a FE model of the cantilever beam [2, 7];
(iii) Identifying the probabilistic properties of the structural behavior.

1.4.3 Results

From the S-shape failure configuration, the apparent contact energy is evaluated as [10]

$$\Gamma = \frac{3}{2} E \frac{h^2 t^3}{l_s^4}, \tag{1.7}$$

where E is the Young's modulus of polysilicon, and l_s is the crack length. The numerical results are compared with the experimental ones for high humidity levels in Fig. 1.8. The results are displayed in terms of the confident ranges of effective adhesive energies, and of the distribution of crack lengths. The comparison shows that the stochastic model-based multi-scale method predicts the uncertainty in the stiction failure of micro cantilever beams with a good accuracy.

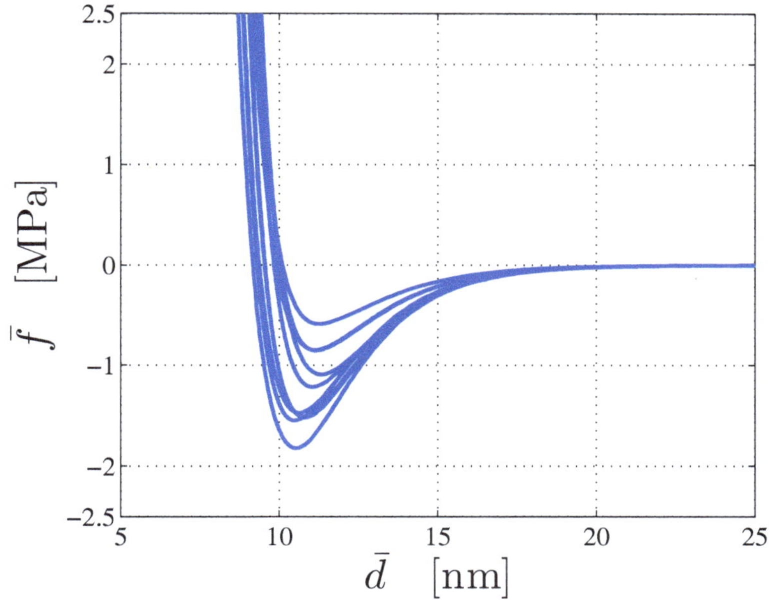

Fig. 1.7 10 generated contact forces at $RH = 85\,\%$ by the stochastic model

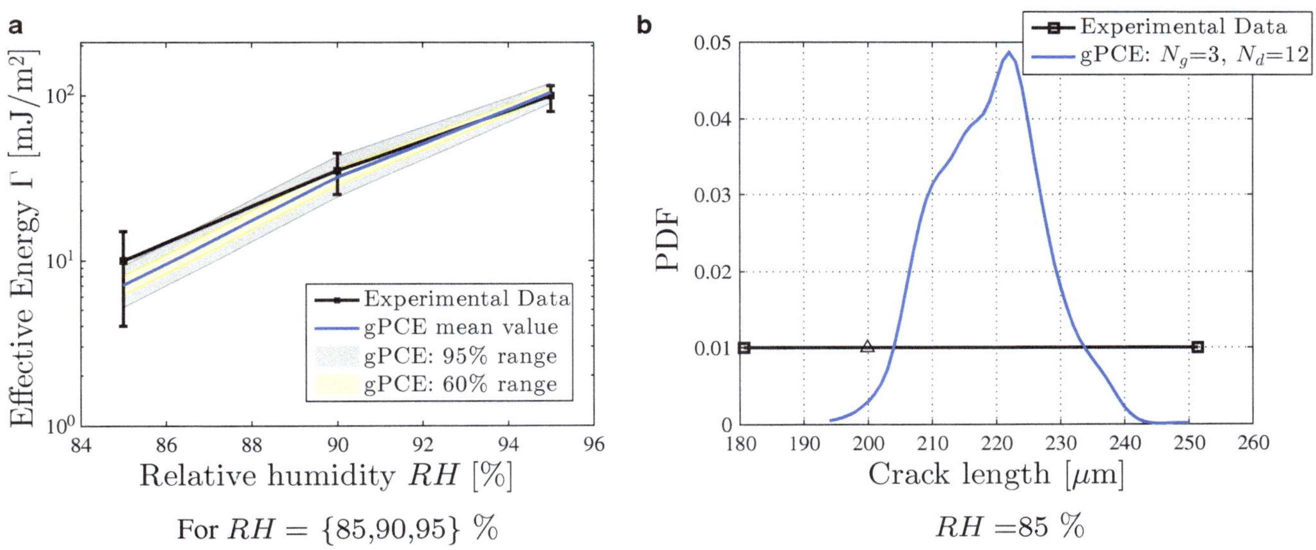

Fig. 1.8 The comparison, in terms of apparent adhesive energies and in terms of distribution of the crack length, between the experimental data, reported in [3], and the numerical results obtained using the stochastic model. (**a**) For $RH = \{85, 90, 95\}\,\%$. (**b**) $RH = 85\,\%$

1.5 Conclusions

The effect of the roughness of contacting surfaces involved in the stiction failure of micro structures induces some uncertainties in the structural behaviors. In order to quantify this uncertainty, a stochastic model-based multi-scale method is developed. The two main ingredients of this method are the multi-scale contact model, and the stochastic model of random apparent contact forces which is used for the uncertainty propagation process. The important advantage of this method is its acceptable computational cost.

When applying the stochastic model-based multi-scale method, the probabilistic behaviors of the humid stiction failure of micro polysilicon cantilever beams in high humidities is well predicted as compared to experimental results.

Acknowledgements The research has been funded by the Walloon Region under the agreement no 1117477 (CT-INT 2011-11-14) in the context of the ERA-NET MNT framework.

References

1. van Spengen, W.M., Puers, R., De Wolf, I.: On the physics of stiction and its impact on the reliability of microstructures. J. Adhes. Sci. Technol. **17**(4), 563–582 (2003)
2. Hoang, T.-V., Wu, L., Paquay, J.-C., Golinval, Arnst, M., Noels, L.: A computational stochastic multi-scale methodology for MEMS structures involving adhesive contact. Tribol. Int. (submitted)
3. DelRio, F.W., Dunn, M.L., de Boer, M.P.: Van der waals and capillary adhesion of polycrystalline silicon micromachined surfaces. In: Bhushan, B. (ed.) Scanning Probe Microscopy in Nanoscience and Nanotechnology. NanoScience and Technology, vol. 3, pp. 363–393. Springer, Berlin (2013). doi:10.1007/978-3-642-25414-7-14. http://dx.doi.org/10.1007/978-3-642-25414-7_14
4. DelRio, F.W., Dunn, M.L.D., de Boer, M.P.: Capillary adhesion model for contacting micromachined surfaces. Scr. Mater. **59**(9), 916–920 (2008). viewpoint set no. 44 "The materials for MEMS". doi:http://dx.doi.org/10.1016/j.scriptamat.2008.02.037. http://www.sciencedirect.com/science/article/pii/S1359646208001668
5. Xue, X., Polycarpou, A.A.: Meniscus model for noncontacting and contacting sphere-on-flat surfaces including elastic-plastic deformation. J. Appl. Phys. **103**(2), (2008). doi:10.1063/1.2830863
6. van Spengen, W.M.: A physical model to describe the distribution of adhesion strength in MEMS, or why one MEMS device sticks and another 'identical' one does not. J. Micromech. Microeng. **25**(12), 125012 (2015). http://stacks.iop.org/0960-1317/25/i=12/a=125012
7. Wu, L., Noels, L., Rochus, V., Pustan, M., Golinval, J.-C.: A micro - macro approach to predict stiction due to surface contact in micro electro-mechanical systems. J. Microelectromech. Syst. **20**(4), 976–990 (2011). doi:10.1109/JMEMS.2011.2153823.
8. Hoang, T.-V., Wu, L., Paquay, S., Obreja, A., Voicu, R., Müller, R., Golinval, J.-C., Noels, L.: A probabilistic model for predicting the uncertainties of the humid stiction phenomenon on hard materials. J. Comput. Appl. Math. **289**, 173–195 (2015)
9. Cai, S., Bhushan, B.: Meniscus and viscous forces during separation of hydrophilic and hydrophobic surfaces with liquid-mediated contacts. Mater. Sci. Eng. R. Rep. **61**(1), 78–106 (2008)
10. de Boer, M.: Capillary adhesion between elastically hard rough surfaces. Exp. Mech. **47**, 171–183 (2007). doi:10.1007/s11340-006-0631-z. http://dx.doi.org/10.1007/s11340-006-0631-z
11. Johnson, K.L.: Contact Mechanics. Cambridge University Press, Cambridge (1987)
12. Hertz, H.: Ueber die berührung fester elastischer körper. J. für die reine und angewandte Math. **92**, 156–171 (1882). http://eudml.org/doc/148490

Chapter 2
Full-Field Identification of Interfaces in Microelectronic Devices

A.P. Ruybalid, J.P.M. Hoefnagels, O. van der Sluis, and M.G.D. Geers

Abstract To improve the integrity of densely stacked multilayers in microelectronic systems, e.g., Light Emitting Diodes (LED), and thereby overcome the currently experienced problems related to interface failure during manufacturing of such devices, accurate identification of interface properties is essential. The behavior of the interface is only measurable through kinematic information from adjacent materials.

The goal of this research is to identify interface parameters by Integrated Digital Image Correlation (IDIC), in which experimental images of a deformation process are correlated by utilizing the mechanical response from finite element (FE) simulations. An interface is herein modeled by cohesive zone (CZ) elements exhibiting constitutive traction-separation laws. The versatility of FE simulations and the kinematic richness of the full-field measurements are thereby exploited.

Comprising an elastic hinge system, a small-scale mechanical test-setup is designed from two 3-axes (XYZ) piezo stages, with which micrometer displacements and realistic interface loading conditions (shear, normal, and mixed-mode loading) can be applied to an LED specimen. This allows to, in a well-controlled manner, mechanically mimic interface delamination that is typically induced during fabrication steps by thermal expansion. This setup and the IDIC method are integrated to identify the CZ parameters of the critical interface of an LED specimen.

Keywords Microelectronics • Interface delamination • Cohesive zone • Full-field measurement • Parameter identification

2.1 Introduction

In order to understand and predict interface behavior in complex, dense material stacks, as used in the microelectronic industry (e.g., in solid state lighting), proper identification of constitutive model parameters is essential. Cohesive Zone (CZ) models are a versatile manner of describing interfaces. Conventionally, CZ-parameters are measured by mechanical tests that are carefully designed so that specific assumptions allow for analyzing material response from limited data (e.g., force-displacement curves). However, the restricted kinematic information may result in inaccurate parameter identification, and it typically calls for more than one test to determine the many different parameters.

To more adequately identify model parameters, a testing rig is required that is capable of mimicking the realistic loading conditions that are responsible for failure of the microelectronic product during application. Moreover, *full-field* identification methods are suggested, in which images of the deformation process are analyzed. At the heart of such a method lies digital image correlation (DIC), which captures full-field kinematics, exhibiting more complete information about material response. Together with the tight integration of finite element (FE) simulation of the conducted mechanical test, images are correlated in order to optimize the CZ-model parameters, using the method of *Integrated Digital Image Correlation*

This proceeding describes the method of Integrated Digital Images Correlation applied to image data from virtual double clamped beam (DCB) experiments and a new testing rig, comprising piezo actuators, for interface testing of microelectronic applications.

A.P. Ruybalid • J.P.M. Hoefnagels (✉) • O. van der Sluis • M.G.D. Geers
Department of Mechanical Engineering, Eindhoven University of Technology, P.O. Box 513, 5600 MB Eindhoven, The Netherlands

L. Starman et al. (eds.), *Micro and Nanomechanics*, Conference Proceedings of the Society for Experimental Mechanics Series,
DOI 10.1007/978-3-319-42228-2_2

2.2 Cohesive Zone Models

Cohesive zone models project all damage mechanisms in and around the crack tip on the interface, leading to a constitutive relation between the traction τ and opening displacement δ between the delaminating material layers. A variety of cohesive zone laws exist [1], of which an exponential implementation is utilized in this research:

$$\tau = G_c \frac{\delta}{\delta_c} \exp\left(-\frac{\delta}{\delta_c}\right), \tag{2.1}$$

where G_c and δ_c are the values for the critical energy release rate and the critical opening displacement, respectively, at which the interface fails. These are the constitutive parameters of interest which must be identified in order to make use of this CZ-model for the understanding and prediction of interface behavior.

2.3 Integrated Digital Image Correlation

In the method of IDIC [2–6], parameter identification is realized by directly integrating the CZ-model simulation with a DIC procedure. In DIC, the residual between images is minimized, assuming conservation of brightness. This is mathematically expressed as:

$$f(\vec{x}, t_0) \approx g\left(\vec{\Phi}(\vec{x}, t)\right) = g \circ \vec{\Phi}(\vec{x}, t), \tag{2.2}$$

$$\vec{\Phi}(\vec{x}, t) = \vec{x} + \vec{U}(\vec{x}, t), \tag{2.3}$$

where f and g are the scalar intensity fields of, e.g., light sensed by a camera sensor, and $\vec{\Phi}$ represents the vector function that maps the pixel coordinates $\vec{x}$ of image f, corresponding to the reference material state, to the pixel coordinates of image g, corresponding to the deformed state. Important to realize is that the mapping function $\vec{\Phi}(\vec{x}, t)$ depends on a displacement field $U(\vec{x}, t)$, which can be approximated by FE-simulation of a model of the experiment: $U(\vec{x}, t) \approx \vec{h}(\vec{x}, t, \theta_i)$, which, in turn, depends on the constitutive model parameters θ_i. The image residual Ψ to be minimized is written as:

$$\Psi = \int_\tau \int_\Omega \frac{1}{2} \left(f(\vec{x}, t_0) - g \circ \vec{\Phi}(\vec{x}, t, \theta_i)\right)^2 \mathrm{d}\vec{x}\mathrm{d}t. \tag{2.4}$$

Hence, optimization of the constitutive parameters θ_i is directly achieved by minimizing this image residual in an iterative Gauss-Newton scheme. A flowchart of the IDIC method is shown in Fig. 2.1.

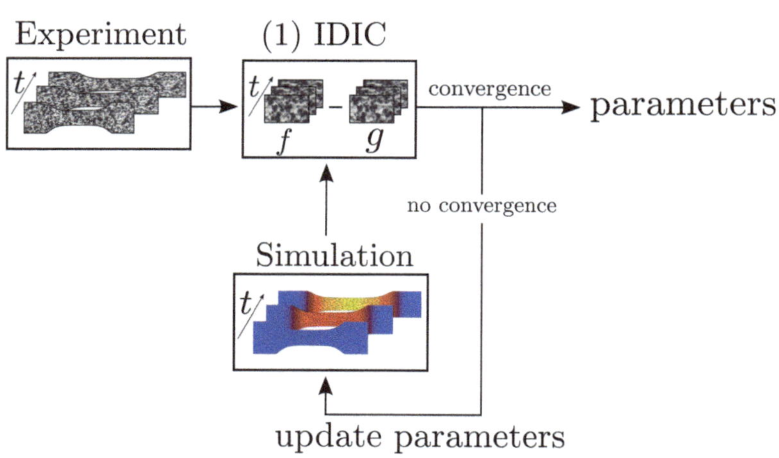

Fig. 2.1 Flowchart of the principle of the full-field identification method of IDIC

Besides kinematic information (in the form of images), loading data, which is typically measured during mechanical testing, is generally required in the identification process for certain parameters [3]. This is mandatory for parameters that directly influence the force level when only Dirichlet boundary conditions are applied in the FE-simulation of the identification methods.

Similar to the displacement and image residuals, a load residual can be defined as:

$$\vec{P}_F = \int_\tau \left(\vec{F}_{\text{exp}}(t) - \vec{F}_{\text{sim}}(t, \theta_i)\right)^2 \text{d}t, \tag{2.5}$$

where $\vec{F}_{\text{exp}}(t)$ is the experimentally measured force, and $\vec{F}_{\text{sim}}(t, \theta_i)$ is the numerically obtained reaction force, which depends on the constitutive parameters θ_i. This objective function is minimized by optimizing the parameters θ_i according to Gauss-Newton optimization, and it can be combined with the minimization processes of the kinematic objective function of equation 2.4.

2.4 Double Clamped Beam Experiment

To prove the principle of IDIC for interface identification and to understand the corresponding difficulties, a virtual DCB experiment is designed that closely matches a true DCB experiment. This means that, instead of taking images of a DCB specimen that is deformed during a mechanical test, deformed images are artificially created with the help of numerical finite element (FE) simulation.

Firstly, an artificial speckle pattern is generated by adding several Gaussian gray value peaks. The resulting reference image (undeformed) can be seen in Fig. 2.2a. The coarseness of the pattern is governed by the widths of the Gaussian peaks, which were tuned so that both fine and coarser speckle features are present in the image, which, respectively, allows for resolving fine kinematics and improves the robustness of the iterative IDIC method. Displacement fields from FE-simulation of a DCB model were used to displace the pixels of the reference image in order to come to a 'deformed' speckle pattern, as seen in Fig. 2.2b. The DCB-model utilizes an elastic constitutive model, described by Young's modulus E and Poisson's ratio ν corresponding to spring steel, for the two beams. Furthermore, a CZ-law is implemented for the glued interface between the beams, which is described by a critical energy release rate parameter G_c and a critical opening displacement parameter δ_c. The boundary conditions applied during the virtual experiment are simple vertical displacements, applied to two (upper and lower) left corner nodes of the two modeled beams.

Another second virtual experiment is performed with a real reference image (coming from the true DCB test) that is virtually deformed, using the same FE-simulated displacement fields. The images used are seen in Fig. 2.2c and d. This additional virtual experiment is done to test the adequacy of a realistic speckle pattern to be used within the IDIC procedure.

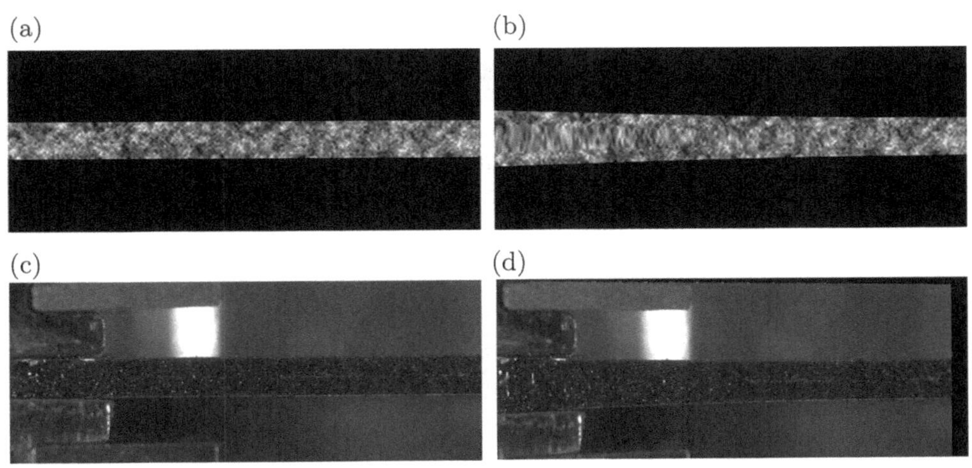

Fig. 2.2 An artificial speckle-pattern and a real, experimental image from a speckle-patterned DCB specimen were virtually deformed to be used within the IDIC procedure to identify the constitutive parameters of the DCB-model, including elastic and CZ parameters. (**a**) Undeformed artificial pattern. (**b**) Virtually deformed artificial pattern. (**c**) Undeformed real pattern. (**d**) Virtually deformed real pattern

2.5 Conclusions and Future Work

The main conclusion drawn from the virtual experiments on the artificially generated speckle patterns are:

- The problem is insensitive to the Poisson's ratio. This parameter is best omitted (locked at a certain value and not optimized) during the iterative identification routine, because it drives the correlation process away from the solution (i.e., it causes divergence).
- Only the mechanical regime of crack growth can be taken into account to solve for the elastic and CZ parameters. This is beneficial, because the elastic regime may be left out of consideration, making the IDIC routine less computationally expensive.
- Even when the initial crack tip location is incorrectly assumed (33 % error of the initial crack tip location), the solution of the CZ-parameters are still accurately identified, with errors well below 1 % (although more erroneous than when the crack tip location was assumed correctly).

The same conclusion hold for the virtual experiment on the realistic speckle pattern, together with the following additional conclusions:

- Similar errors for the elastic, and cohesive zone parameters remain after convergence, as for the virtual experiment with the artificial speckle pattern.
- The actual boundary conditions (BC's) measured from the experimental data are applied in the virtual experiment. When the simulation used within the IDIC routine also uses these same BC's, accurate results are obtained, with errors for the parameters remaining below 1 %.
- When 3 % of the dynamic range of gray levels of noise is added to the images, convergence to the solutions is still accurately reached.
- Gaussian blurring of images (to decrease the influence of image noise) does not contribute to higher accuracy of the parameters.
- When the interface is made heterogeneous in the virtual experiment (meaning that over the interface length, the cohesive zone parameters vary), and the simulation within the IDIC routine uses a homogeneous description of the interface, the same resulting image residual is obtained as for the homogeneous experiment. The parameters converge to values that correspond to the average values of the varying parameters of the heterogeneous interface of the virtual experiment. This indicates that even heterogeneous interfaces (which are the most-likely present in realistic applications) can be identified with this identification method.

In order to identify interface parameters with IDIC in realistic, microelectronic applications, a testing rig is required that can handle the small-scale specimens and apply realistic loading conditions. For this purpose, two stacks of x-, y-, and z-axis piezo actuators will together form a testing rig that can load a microelectronic specimen under tensile and shear loading conditions. The actuators can each generate 10 N of force and can displace 200 μ m in order to realize delamination of material layers at the most critical interfaces of the material stack.

The two piezo stacks, as seen in Fig. 2.3, are each controlled by a closed-loop controller, which allows for accurate positioning and force measurement in each direction.

A clamping body is used to clamp a specimen and transfer the load in order to realize mechanical deformation and delamination between material layers.

Due to the vacuum compatibility and the compact size of the systems, in-situ SEM investigation is possible which allows for observing deformation of the microscopically thin material layers, providing the images that are eventually used for the full-field identification technique of IDIC.

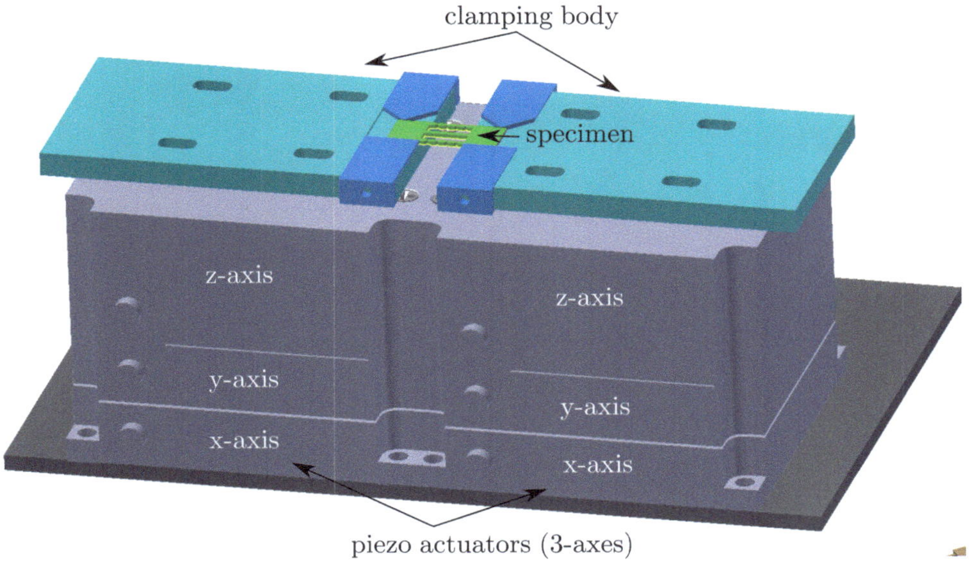

Fig. 2.3 Computer aided designed model of the piezo actuator stacks that will be used to load microelectronic devices under tensile and shear loading conditions

References

1. Kolluri, M.: An in-situ experimental-numerical approach for interface delamination characterization. Ph.D.-thesis, Eindhoven University of Technology, 2011
2. Leclerc, H., Périé, J.N., Roux, S., Hild, F.: Integrated digital image correlation for the identification of mechanical properties. In: Computer Vision/Computer Graphics Collaboration Techniques. Lecture Notes in Computer Science (including subseries Lecture Notes in Artificial Intelligence and Lecture Notes in Bioinformatics), Lecture Notes in Computer Science, vol. 5496, pp. 161–171. Springer, Berlin (2009)
3. Réthoré, J., Muhibullah, Elguedj, T., Coret, M., Chaudet, P., Combescure, A.: Robust identification of elasto-plastic constitutive law parameters from digital images using 3D kinematics. Int. J. Solids Struct. **50**(1), 73–85 (2013)
4. Réthoré, J., Roux, S., Hild, F.: An extended and integrated digital image correlation technique applied to the analysis of fractured samples. Eur. J. Comput. Mech. **18**, 285–306 (2009)
5. Roux, S., Hild, F.: Stress intensity factor measurements from digital image correlation: Post-processing and integrated approaches. Int. J. Fract. **140**(1–4), 141–157 (2006)
6. Ruybalid, A.P., Hoefnagels, J.P.M., van der Sluis, O., Geers, M.G.D.: Comparison of the identification performance of conventional FEM updating and integrated DIC. Int. J. Numer. Methods Eng. (2015)

Chapter 3
Experimental Study of Microstructure and Mechanical Property of Cu30Zn6Al Alloy

Haohan Li and Qizhen Li

Abstract This work focused on the synthesis, and microstructural and mechanical characterizations of copper-30 %zinc-6 %aluminum (Cu30Zn6Al) alloy. The alloy was first synthesized using an induction furnace, then homogenized for different durations, and heated and quenched using one-step and two-step methods. Optical microscope and scanning electron microscope were employed for the microstructure characterization. Microhardness testing was performed to obtain the Vickers hardness of the alloy after the quenching processes. The increase of homogenization duration from 2 to 12 h increases the average grain size from about 50 μm to about 100 μm when the samples were homogenized at 500 °C. The microhardness of the alloy varied in the range of about 40 HV to about 75 HV after experiencing the quenching processes.

Keywords Copper-zinc-aluminum alloy • Casting • Heat treatment • Quenching • Microhardness

3.1 Introduction

Cu alloys are broadly utilized in various fields such as MEMS, electrical appliances, architecture, seawater, tubes, pipes and fittings because they have high electric conductivity and good machinability, and they are also corrosion resistant and can stand the harsh seawater environment [1]. Additionally, Cu alloys are less expensive comparing with other nonferrous alloys. There is extensive research effort on improving the mechanical property of Cu through adding alloying elements such as Zn and Al and performing severe plastic deformation (SPD) on the alloys [2–9]. Instead of performing SPD processing, heat treatment can also change the microstructure and mechanical property of CuZnAl alloys. In this study, Zn and Al were added to Cu to produce Cu30Zn6Al alloy and various heat treatment processes were performed to explore their effect on microhardness of the Cu30Zn6Al alloy.

3.2 Experimental Methods

The Cu30Zn6Al samples were prepared using 99.9 % pure Cu, 99.99 % pure Zn, and 99.99 % pure Al in an induction furnace. Three groups of heat treatment processes were performed on the samples as listed in Table 3.1. Group No. 1 can be referred as a homogenization process, and the samples were heated at 500 °C for 2, 4, 6, and 12 h respectively and then cooled in water at room temperature. For Group No. 2, the samples were heated at 900 °C for 5, 10, 15, 20, 25, and 30 min respectively and then followed by a one-step quenching to water at room temperature directly. For Group No. 3, the samples were heated at 900 °C for 5, 10, 15, 20, 25, and 30 min respectively and then experienced a two-step quenching. During the two-step quenching, the samples were first quenched to the boiling water for 30 min and then to water at room temperature.

H. Li
University of Nevada-Reno, Reno, NV 89557, USA

Q. Li (✉)
University of Nevada-Reno, Reno, NV 89557, USA

Washington State University, Pullman, WA 99164, USA
e-mail: qizhen.li@wsu.edu

L. Starman et al. (eds.), *Micro and Nanomechanics*, Conference Proceedings of the Society for Experimental Mechanics Series,
DOI 10.1007/978-3-319-42228-2_3

Table 3.1 Three groups of heat treatment processes performed on the Cu30Zn6Al samples

Group no.	Heating temperature (°C)	Heating duration	Cooling
1	500	2, 4, 6, and 12 h	Water at room temperature
2	900	5, 10, 15, 20, 25, and 30 min	One-step quenching: to water at room temperature
3	900	5, 10, 15, 20, 25, and 30 min	Two-step quenching: to the boiling water for 30 min and then to water at room temperature

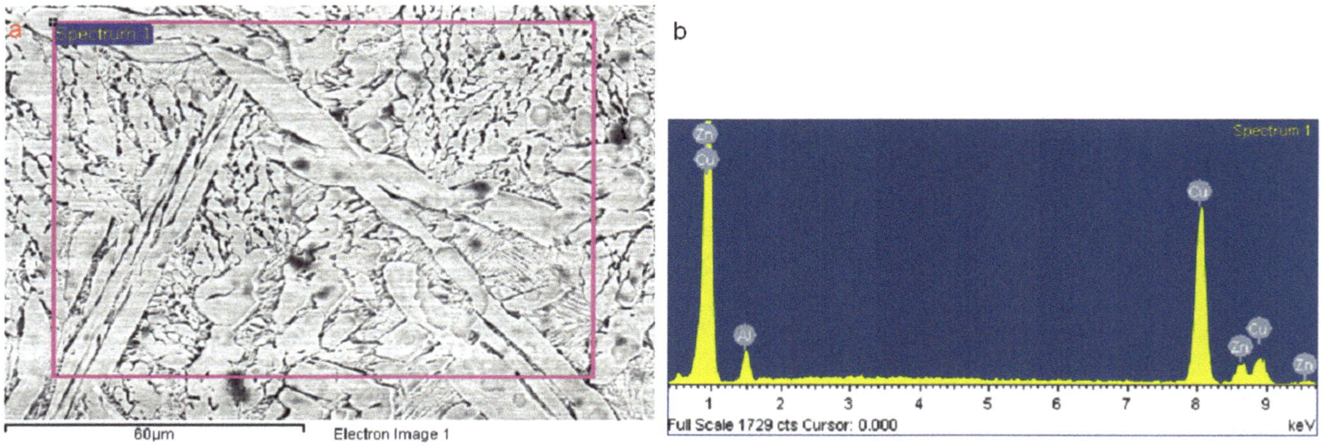

Fig. 3.1 (**a**) SEM image and (**b**) EDX analysis of Cu30Zn6Al alloy

The processed samples were ground, polished, and etched using the NH_4OH and H_2O_2 solution for microstructure observation. Microstructural characterization was carried out by visual light microscope (VLM) and scanning electron microscope (SEM). The elemental analysis was conducted by the energy dispersive x-ray spectroscopy (EDX). Ten microhardness measurements were collected and averaged for each of the samples experienced one-step quenching and two-step quenching respectively.

3.3 Results and Discussion

Figure 3.1a shows the SEM image of a Cu30Zn6Al sample and the region inside the rectangle was scanned to collect the EDX data as shown in Fig. 3.1b. The data indicate the existence of Cu, Zn, and Al in the sample. Figure 3.2 reports the optical micrographs of the as-cast Cu30Zn6Al alloy.

The optical micrographs for the homogenized samples are shown in Fig. 3.3. With the increase of the homogenization duration from 2 to 12 h, the grain size of the corresponding Cu30Zn6Al samples increases from about 50 μm to about 100 μm.

Figures 3.4a–f reports the optical micrographs of Cu30Zn6Al samples heated at 900 °C for 5, 10, 15, 20, 25, and 30 min respectively and then quenched to water at room temperature. The micrographs show the martensite structure of the samples and the average martensite plate sizes are about 4.2, 4.5, 4.9, 5.2, 5.4 and 5.5 μm for the six heating durations (i.e. 5, 10, 15, 20, 25, and 30 min) in Fig. 3.4 respectively. The plate size increases slightly with the increase of heating duration at 900°C.

Figures 3.5a–f reports the optical micrographs of Cu30Zn6Al samples heated at 900 °C for 5, 10, 15, 20, 25, and 30 min respectively and then experienced a two-step quenching. The micrographs show the martensite structure of the samples and the average martensite plate sizes are about 3.3, 3.8, 5.2, 5.3, 6.1, and 6.4 μm for the six heating durations (5, 10, 15, 20, 25, and 30 min) in Fig. 3.5 respectively. The plate size increases slightly with the increase of heating duration. Additionally, there is no significant difference in the plate size between the samples directly quenched to water at room temperature and those quenched to the boiling water for 30 min and then to water at room temperature.

Figure 3.6 shows the hardness data for the samples heated at 900 °C for 5, 10, 15, 20, 25, and 30 min respectively and then quenched. For both quenching routes, the samples have no significant difference in hardness values when they were heated at 900 °C for 5 min. When the heating duration is longer than 5 min, the sample quenched to water at room temperature has a higher hardness than that experienced two-step quenching for each heating duration. This is possibly due to the continuously heating at 100 °C for 30 min when a sample experiences the two-step quenching. The data in Fig. 3.6 also show that the

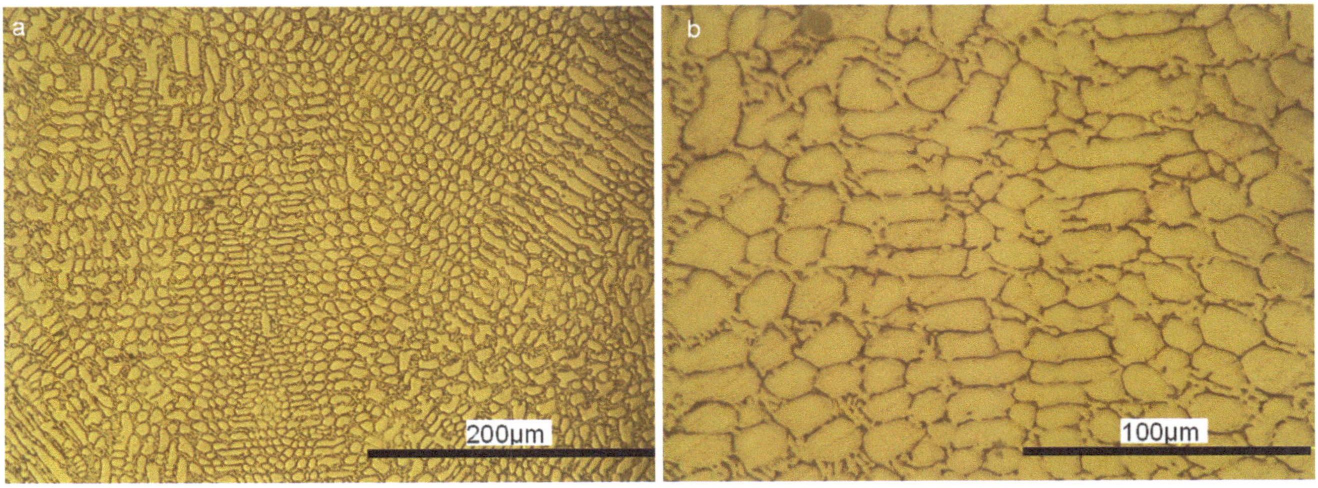

Fig. 3.2 Optical micrograph of Cu30Zn6Al alloy at (**a**) low magnification and (**b**) high magnification

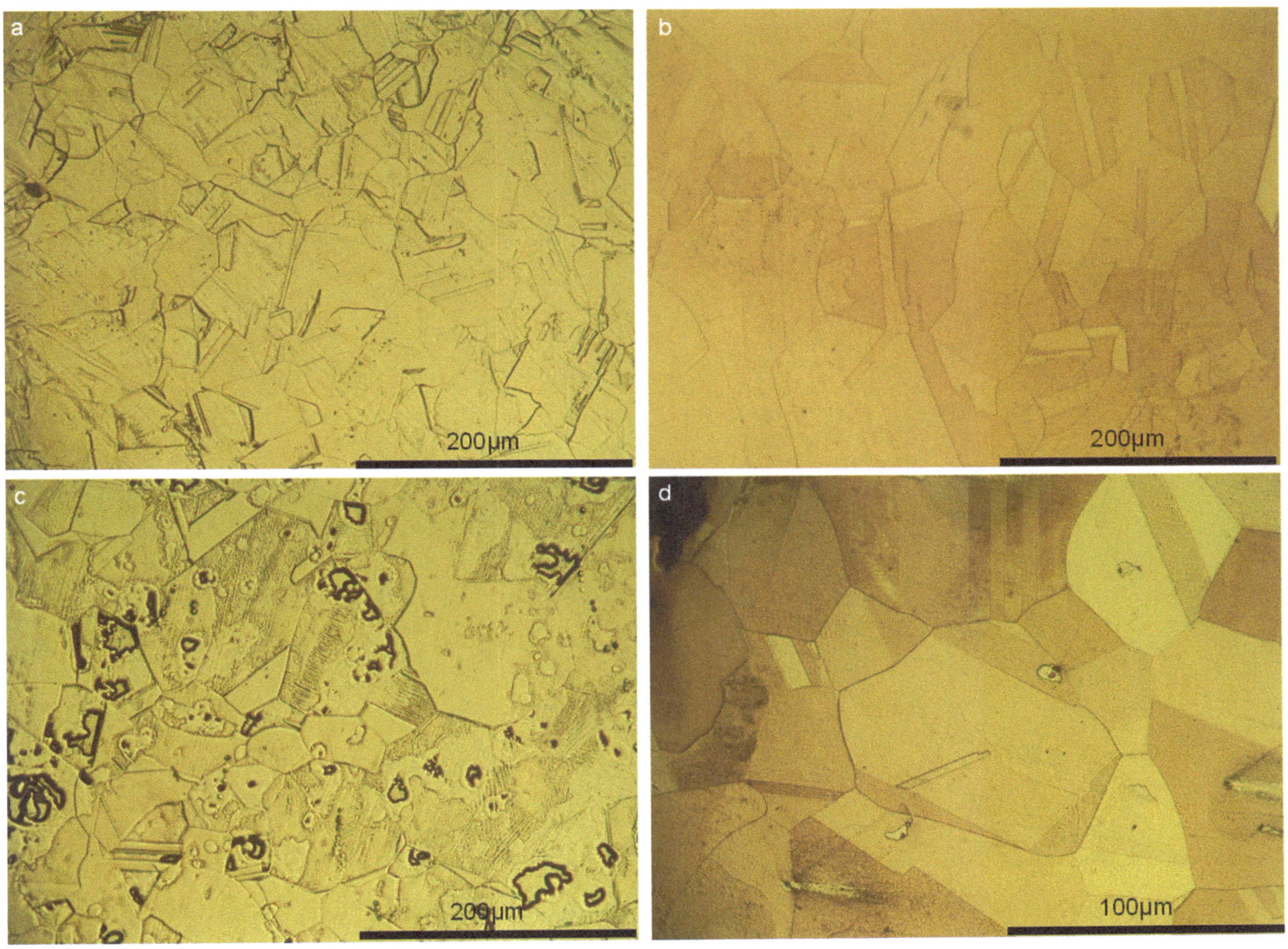

Fig. 3.3 Optical micrographs of Cu30Zn6Al alloy after being homogenized at 500 °C for (**a**) 2 h, (**b**) 4 h, (**c**) 6 h, and (**d**) 12 h respectively

hardness value decreases with the increase of heating duration from 5 to 15 min, increases from 15 to 20 min, and decreases again from 20 to 30 min for both one-step quenching and two-step quenching. The hardness data fall in the range of about 40 HV to about 75 HV.

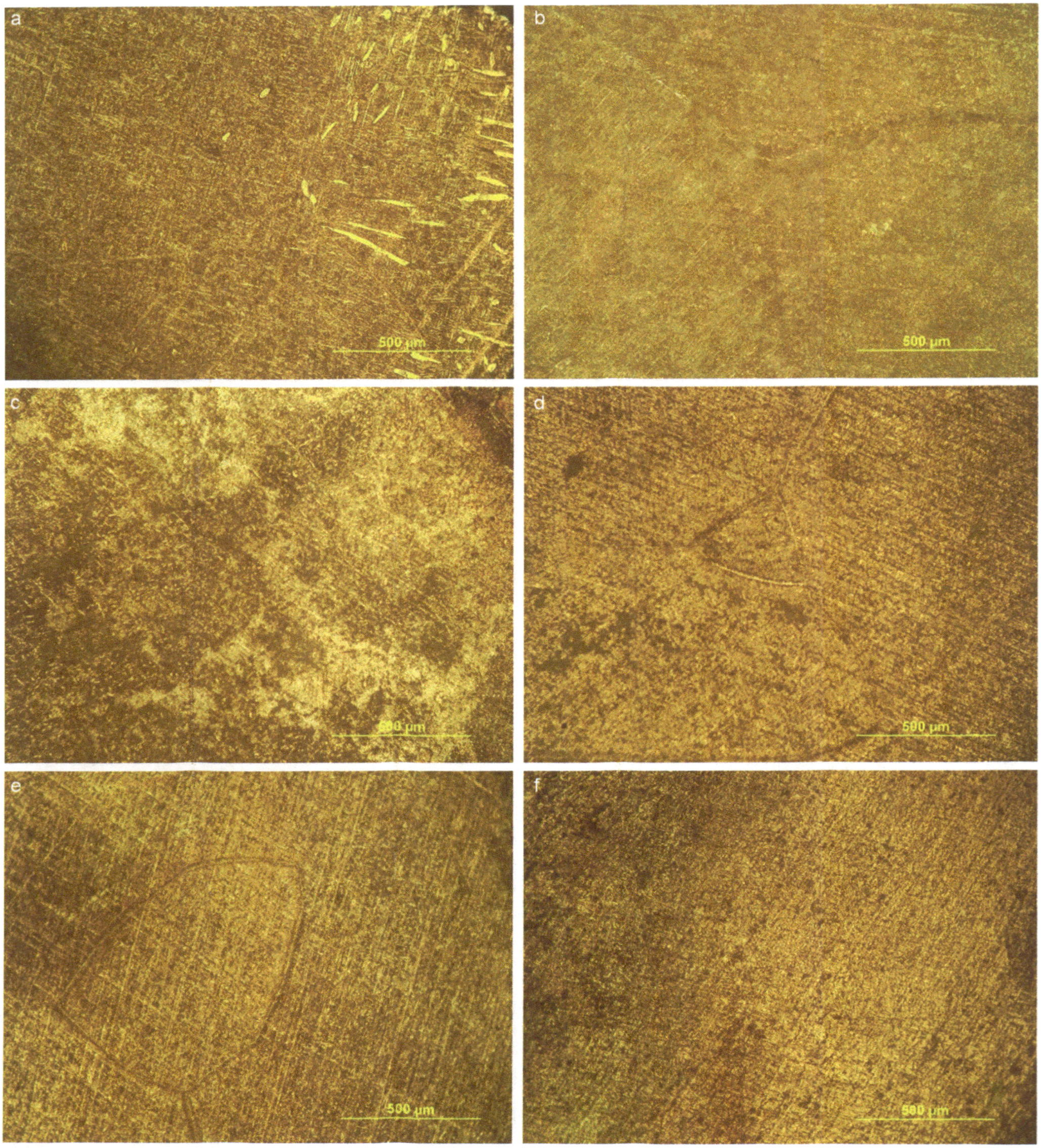

Fig. 3.4 Optical micrographs of Cu30Zn6Al alloy after being heated at 900 °C for (**a**) 5 min, (**b**) 10 min, (**c**) 15 min, (**d**) 20 min, (**e**) 25 min, and (**f**) 30 min respectively and then followed by the one-step quenching to water at room temperature directly

3.4 Conclusion

The Cu30Zn6Al samples were prepared using an induction furnace and then heat treated through three groups of processes. Group No. 1 was homogenized at 500 °C for a range of durations (i.e. 2, 4, 6, and 12 h) and then cooled to water at room temperature. Group No. 2 was heated at 900 °C for a range of durations (i.e. 5, 10, 15, 20, 25, and 30 min) and followed

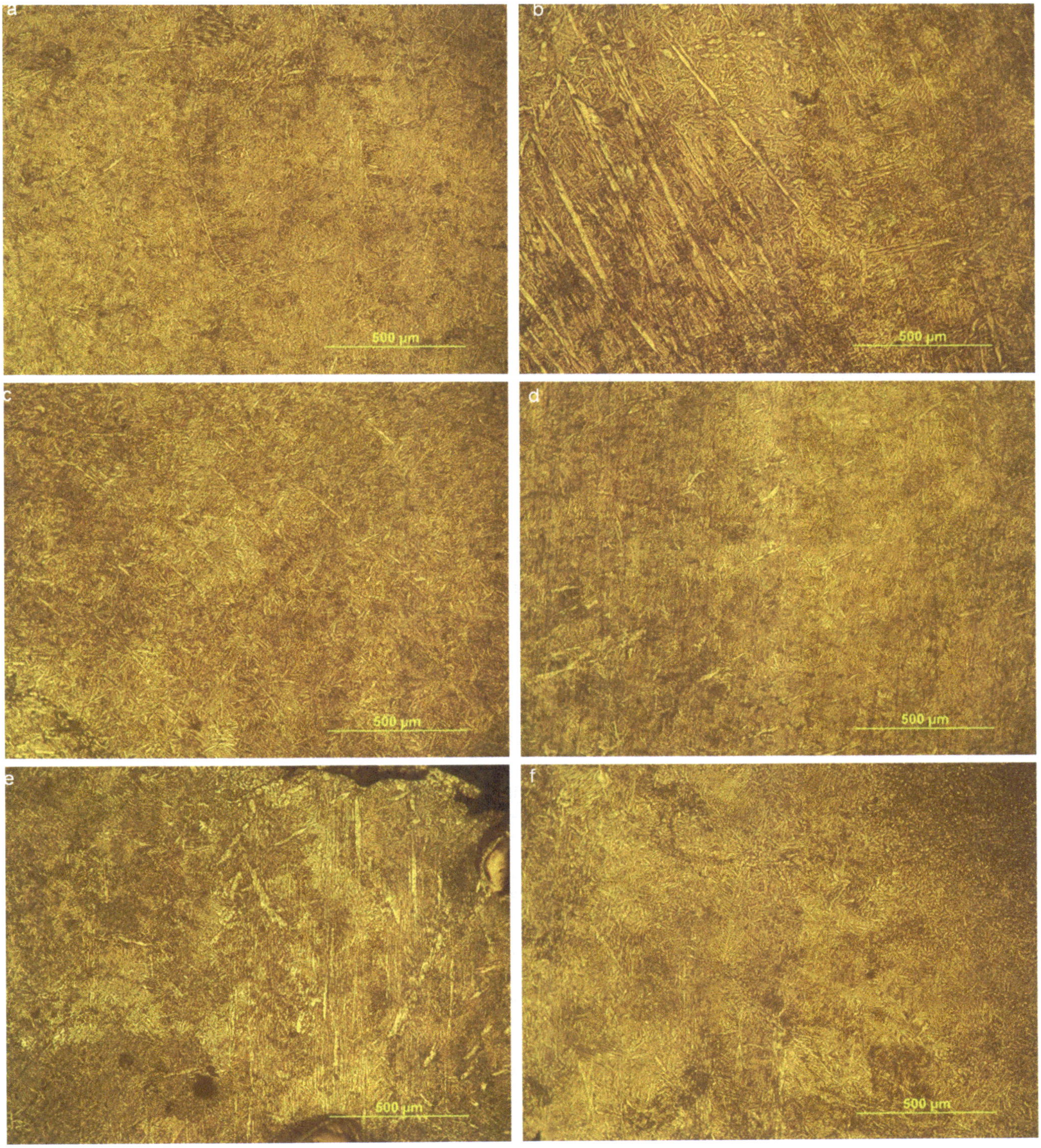

Fig. 3.5 Optical micrographs of Cu30Zn6Al alloy after being heated at 900 °C for (**a**) 5 min, (**b**) 10 min, (**c**) 15 min, (**d**) 20 min, (**e**) 25 min, and (**f**) 30 min respectively and then experienced the two-step quenching

by the one-step quenching. Group No. 3 was heated at 900 °C for a range of durations (i.e. 5, 10, 15, 20, 25, and 30 min) and followed by the two-step quenching. The one-step quenching is quenching in water at room temperature directly. The two-step quenching is quenching to the boiling water for 30 min and then to water at room temperature. The increase of homogenization duration from 2 to 12 h increases the average grain size from about 50 μm to about 100 μm. The average

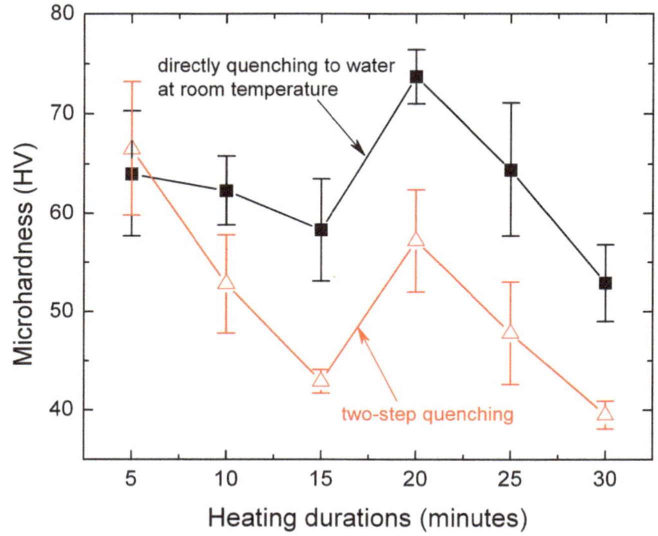

Fig. 3.6 Microhardness of the Cu30Zn6Al samples heated at 900 °C for 5, 10, 15, 20, 25, and 30 min respectively and then quenched through one-step quenching and two-step quenching respectively

martensite plate sizes are in the range of 4.2 to 5.5 μm for the samples experienced one-step quenching processing, and in the range of 3.3 to 6.4 μm for the samples experienced two-step quenching processing. The microhardness of the alloys varied in the range of about 40 HV to about 75 HV after experiencing one-step quenching and two-step quenching. In general, the samples experienced one-step quenching have higher hardness values than those experienced two-step quenching when they experienced the same pre-quenching heat treatment.

References

1. Davis, J.R.: ASM Specialty Handbook: Copper and Copper Alloys. ASM International, Materials Park, OH (2001)
2. Wang, Y.B., Liao, X.Z., Zhao, Y.H., Lavernia, E.J., Ringer, S.P., Horita, Z., Langdon, T.G., Zhu, Y.T.: The role of stacking faults and twin boundaries in grain refinement of a Cu–Zn alloy processed by high-pressure torsion. Mater. Sci. Eng. A **527**, 4959–4966 (2010)
3. Bahmanpour, H., Youssef, K.M., Scattergood, R.O., Koch, C.C.: Mechanical behavior of bulk nanocrystalline copper alloys produced by high energy ball milling. J. Mater. Sci. **46**, 6316–6322 (2011)
4. Zhang, Z.J., An, X.H., Zhang, P., Yang, M.X., Yang, G., Wu, S.D., Zhang, Z.F.: Effects of dislocation slip mode on high-cycle fatigue behaviors of ultrafine-grained Cu–Zn alloy processed by equal-channel angular pressing. Scr. Mater. **68**, 389–392 (2013)
5. Jiang, H., Zhu, Y.T., Butt, D.P., Alexandrov, I.V., Lowe, T.C.: Microstructural evolution, microhardness and thermal stability of HPT-processed Cu. Mater. Sci. Eng. A **290**, 128–138 (2000)
6. Wang, Y.M., Ma, E., Chen, M.W.: Enhanced tensile ductility and toughness in nanostructured Cu. Appl. Phys. Lett. **80**, 2395–2397 (2002)
7. Chen, J., Lu, L., Lu, K.: Hardness and strain rate sensitivity of nanocrystalline Cu. Scr. Mater. **54**, 1913–1918 (2006)
8. Cao, W.Q., Gu, C.F., Pereloma, E.V., Davies, C.H.J.: Stored energy, vacancies and thermal stability of ultra-fine grained copper. Mater. Sci. Eng. A **492**, 74–79 (2008)
9. Hosseini, S.A., Manesh, H.D.: High-strength, high-conductivity ultra-fine grains commercial pure copper produced by ARB process. Mater. Des. **30**, 2911–2918 (2009)

Chapter 4
Boundary Mechanics in Lath Martensite, Studied by Uni-Axial Micro-Tensile Tests

J.P.M. Hoefnagels, C. Du, and M.G.D. Geers

Abstract Lath martensite is the key constituent in advance steels that provides the overall strength. Martensite is known as hard and brittle, but recent evidence shows significant plasticity before fracture. The exactly reason is undisclosed but should relate to the underlying lath microstructure. Therefore, we studied the influence of sub-block and block boundaries on martensite plasticity through uni-axial tensile testing of individual micro-constituents, i.e. single block specimens and specimens with a single through-thickness block boundary parallel, perpendicular and at 45° to loading.

A unique micro-tensile methodology was developed included micro-specimen fabrication with minimal FIB damage, EBSD at top and bottom surfaces, a home-built highly-sensitive uni-axial tensile tester, and in situ microscopic slip trace analysis. Interestingly, all specimens showed extensive plasticity before fracture and no cleavage, however, strong differences are observed. Detailed analysis of the rich experimental data shows that not only the block but also the sub-block boundaries show boundary strengthening following a Hall–Petch relation, in the case that the easiest slip systems are crossed by these boundaries. However, for boundaries oriented under 45°, often easy glide is observed along the boundary reducing the strength, possibly caused by retained austenite films at the boundaries. TEM analysis of the boundary structure is ongoing.

Keywords Ductile damage • Lath martensite • Micromechanics • Damage characterization • In situ mechanical testing

4.1 Introduction

Lath martensite, as the major morphology of martensite, has significant industrial importance because it is the basic structure of high strength steels, such as dual-phase steel, transformation-induced plasticity steel, quench-partitioning steel. The reason behind is that martensite is the constituent that brings the strength to the alloy. Since decades, the research about the strengthening mechanisms has been carried out. The strengthening mechanisms have been figured out to be (1) forest dislocation hardening [1, 2], (2) solid solution hardening by alloying elements [2], (3) precipitation strengthening, e.g., by carbides [2, 3], and most importantly (4) substructure boundary strengthening [3–10]. The hierarchical structure, which shows substructures of packets, blocks and sub-blocks from one prior austenite grain, gives lath martensite huge amount of internal boundaries [1, 2]. These boundaries, acting as barriers of dislocation movements constitute the so-called 'morphological constituents strengthening' [3]. It has been proposed that among all the internal boundaries, the block boundary is the most effective boundary [4–6]. This would mean that the blocks size can be regarded as the effective grain size in lath martensite.

The research on the effect of substructures of martensite in literature, which concentrates on the strengthening, can be grouped in the following categories: macro-sized and micro-sized specimens. In [4, 5], macro-sized specimen have been tested, and the relationship between 0.2 % yielding strength and block size was found to follow the Hall–Petch equation. From this measurement it was suggested that the block size is the effective grain size. However, this is an indirect investigation, from which it is impossible to directly observe the individual block boundaries and observe the plastic behavior locally. Ohmura has carried out a series of nano- and micro-indentation tests on lath martensite and concluded that block structure increase the hardness of martensite, using the ratio between nano-hardness and micro-hardness [3, 7]. It was concluded that a smaller ratio would indicate a larger substructure-size effect. However, this conclusion remains ambiguous because the contribution from the high micro-indentation could come as well from the other boundaries, notably the packet boundaries. Micro-bending tests have been carried out by Shibata [5, 8], in which micro-sized bending specimen with and without block boundaries were investigated. It was proposed that dislocations pile up at block boundaries and subsequently propagate across

J.P.M. Hoefnagels (✉) • C. Du • M.G.D. Geers
Department of Mechanical Engineering, Eindhoven University of Technology, P.O. Box 513, 5600 MB Eindhoven, The Netherlands
e-mail: j.p.m.hoefnagels@tue.nl

L. Starman et al. (eds.), *Micro and Nanomechanics*, Conference Proceedings of the Society for Experimental Mechanics Series,
DOI 10.1007/978-3-319-42228-2_4

the block boundary. The fact that no local slip bands appears for the specimen with block boundary was seen as evidence that block boundary restricts dislocation motion and acts as the most effective grain boundary in lath martensite. However, due to the complicated stress states on the bending sample, Schmid analysis was almost impossible. Lath martensite was also tested with micro-pillar compression test by Ghassemi-Armaki [9, 10]. Single block specimens were seen to show perfect elasto-plastic behavior, whereas multiple block specimen show significant strain hardening. In both references [9, 10], however, the microstructure of the multi-block specimens was complicated, perhaps even including one or more packet boundaries, therefore, it is difficult to determine whether the hardening is due to the block boundary or due to the packet boundary. Finally, very interesting micro-tension tests on lath martensite have been conducted by Mine [11], in which specimens from single packet, multi-packets and prior austenite grain were tested. For the single packet specimens, however, only one configuration of block boundary was conducted, which makes it in-sufficient to discuss the effect of block boundary.

In general, all the researches mentioned above was focused on the strengthening effects of boundaries of lath martensite, which is attributed to the fact that dislocations need to cross these boundaries during the plastic deformation. However, in engineering materials like DP steel, martensite islands are often small and do not always contains all the variants. In this case, dislocation may not have to cross the boundary to realize local deformation if the most favored slip system is aligned with the boundaries [12]. Also in fully martensitic steel, plasticity was also observed to occur parallel to the (tilted) lath boundaries, although at large strains slip cuts across boundaries [13, 14].

Thus, for both multi-phase steels, which contain lath martensite as one of the constituents, and fully martensitic steels, the dislocation may not need to move across the boundaries if the boundary is properly aligned. Therefore, the role of boundary could be very different from the strengthening mechanism suggested in the literature. The goal of this work is to study the role of boundaries in the micro-mechanics of lath martensite, both in configurations where the dislocations do and do not need to propagate across the boundaries.

In this study, we perform in situ uniaxial micro-tensile tests, using a home-built nano-force tensile tester [15], of lath martensite specimens from single packets consisting of different types of boundaries with different angles with respect to the loading direction. It will be shown that both block boundary and sub-block boundary play a very important role in the strengthening of lath martensite when boundaries are approximately parallel to the loading direction, however, that a new deformation micro-mechanisms other than dislocation slip is observed in those cases where the boundaries are titled at approximately 45° to the loading direction.

4.2 Experiment

The material has a chemical content of 0.092C-1.68Mn-0.24Si-0.57Cr. It was first homogenized at austenite temperature in an vacuum furnace to obtain large grains, then followed by water quenching. The experiment involves fabrication of a wedge shape from a lath martensite sheet and careful selection of the specimen location based on large EBSD maps, focused ion beam (FIB) fabrication of the micro-tensile specimens combined with detailed top- and bottom-side EBSD analysis of each specimen, and uniaxial tensile tests with highly accurate specimen alignment and force- and displacement measurement. Figure 4.1 shows the experimental setup. Two configurations of samples are prepared: (1) Samples with boundaries tilted to the loading direction and (2) Samples with boundaries parallel to the loading direction.

4.3 Results and Discussion

4.3.1 Samples with Boundaries Parallel to the Loading Direction

Figure 4.2 shows two multi-block samples with block boundary parallel to the loading direction [16]. SEM images of the fractured samples are shown together with inverse pole figures. The red lines indicate the fracture surface. The zig-zag fracture surfaces correspond well with the positions of the block boundaries. This suggests that the fracture surface cannot be continuous across the block boundaries. In other words, fracture surface cannot propagate across block boundaries.

In [17], also a number of three single block samples with sub-block boundaries approximately parallel to the loading direction are discussed. For all the three samples, the traces of the fracture surface match well with the {110} plane with the highest Schmid factor, which means that the slip system in this lath martensite is {110}<111> family. The fracture surface and the slip traces seem to be continuous across the sub-block boundary and no difference could be seen in the two sub-blocks due to the small miss-orientation. Schmidt factor analysis also indicates a very close slip trace of two sub-blocks in single block specimens and the slip directions are almost the same. A further analysis of these results is given in [16].

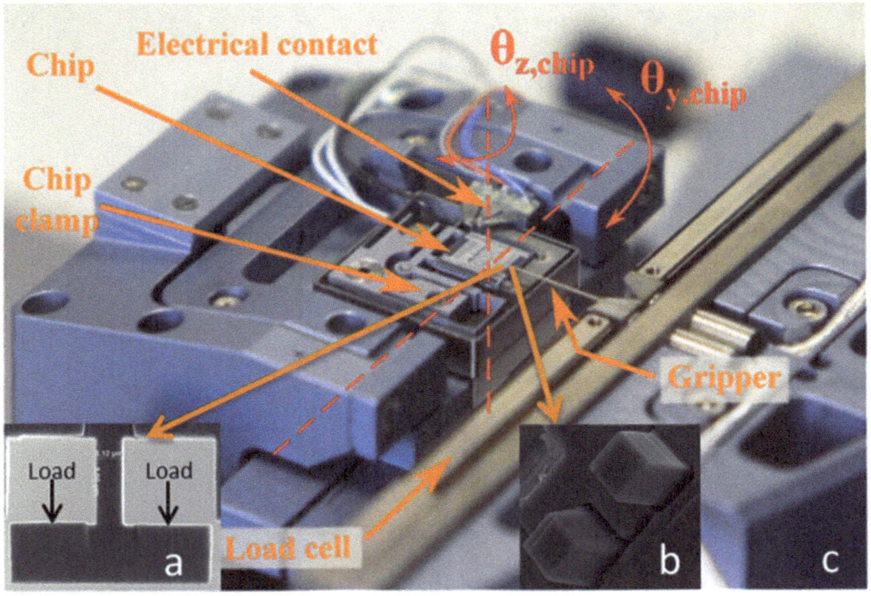

Fig. 4.1 (**a**) An example of micro-tensile specimen, which has a gauge dimension of $9 \times 3 \times 2.5\ \mu m^3$. The loading is applied at the shoulder part symmetrically. (**b**) The two-teeth gripper that is used to apply the loading. (**c**) A close view of the experimental setup (only the specimen mounting part and the load cell part). The positions of the micro-specimens and the gripper are indicated with *arrows*. The gripper is at the top of a loading arm, which is connected to a load cell that is moved by piezo material with high accuracy. Detailed information about the whole testing procedure can be found in [17]. Figure reproduced from [16]

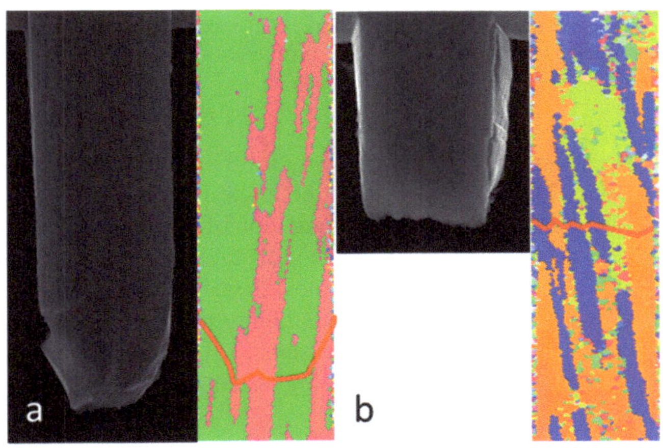

Fig. 4.2 (**a**, **b**) The fracture surface (*left*) and EBSD map (*right*) of two multi-block samples with block boundaries parallel to the loading direction. The length of undeformed samples is 9 μm. *Red curves mark* the fracture traces in the EBSD maps. Figure reproduced from [16]

4.3.2 Samples with Boundaries Tilted to the Loading Direction

Figure 4.3 shows two samples with boundaries tilted to the loading direction. Figures 4.3a–d shows a single-packet, multi-block specimen with boundary tilted to the loading direction. Inverse pole figure and pole figure are shown together with SEM pictures before and after deformation. It is obvious that sliding has happened during the plastic deformation. Analysis of the inverse pole figure in Fig. 4.3a showed that the sliding occurred at the boundary between the two block (blue and red). The red circles in Fig. 4.3d mark, for each of the two blocks, the slip system with the highest Schmid factor. Schmid analysis shows that crystallographic slip could happen with high possibility since the Schmid factor is high in both blocks. However, sliding occurred at the boundary dominantly although it is not straightforward to identify the fracture surface. This suggests that with this configuration, sliding is easier than crystallographic slip.

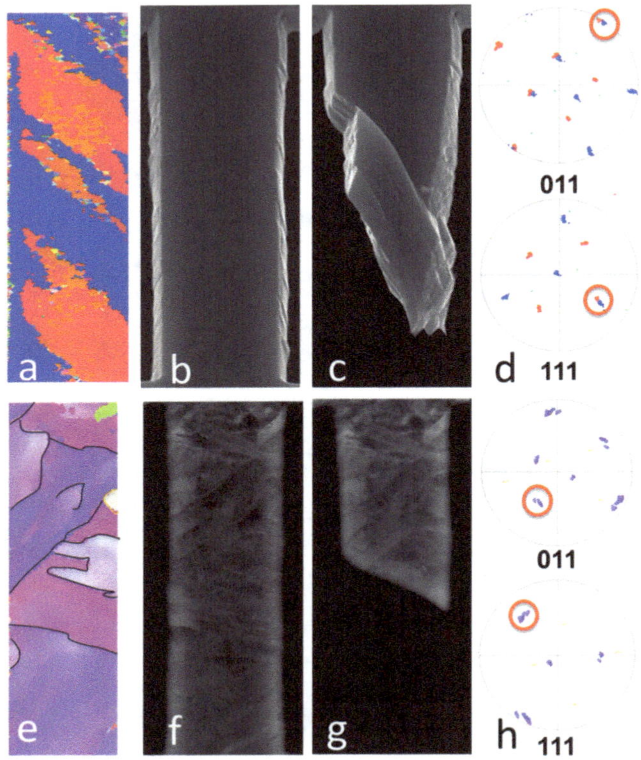

Fig. 4.3 Two lath martensite micro-tensile specimens with boundaries tilted to the loading direction. (**a**, **e**) Inverse pole figure; (**b**, **c**, **f**, **g**) SEM picture of the specimen before/after deformation; (**d**, **h**) {110} and {111} pole figures with *red circles* indicating the favored slip system. Figure reproduced from [18]

Figures 4.3e–h shows a single-block sample with (almost) only sub-block boundaries. The sub-block boundaries marked by black lines in the inverse pole figure. The slip system with the highest Schmid factor is again indicated with red circles in the pole figure. Considering the microstructure with its configuration of sub-block boundaries, one might expect slip to be activated at many positions across the specimen length since the probability for slip is equal at all these positions. However, the fracture surface again lies exactly at one of the sub-block boundary. This strongly suggests that when the boundary is tilted, plastic deformation is realized more easily by boundary sliding than crystallographic slip. More analysis is needed, however, to confirm the presence and nature of this sliding mechanism. To this end, more experiments on single-packet and single-block specimens, with tilted boundaries, are currently being tested. These results will be published elsewhere [18].

4.4 Overview of All Tested Specimens

In Fig. 4.4, the critical resolved shear stress, τ_{CRSS}, is plotted as a function of the square root of the number of boundaries. The critical stress, τ_{CRSS}, is calculated by multiplying the fracture stress by the highest Schmidt factor of the block with largest volume fraction. Three series of data points are plotted. The blue ones indicate single-packet samples with block boundaries approximately parallel to the loading direction, while the green ones indicate single-block samples with sub-block boundaries approximately parallel to the loading direction. The purple ones, all with number of boundaries to be 0, are the samples with boundaries titled at approximately 45° to the loading direction. For both types of parallel-boundary specimens, a linear trends is observed between τ_{CRSS} and the square root of the number of boundaries, which indicates that a Hall–Petch-like relationship is active. It is seen that the slope of the linear fir is slightly larger for the block boundary strengthening compared to the one for the sub-block boundary strengthening. This suggests that the strengthening effect of block boundary is slightly higher than that of the sub-block boundaries, which corresponds to the fact that the fracture was observed to propagate through sub-block boundaries, but not block boundaries. The data points of tilted boundaries shows scattering as well, which may be attributed to the types of boundaries, the shape and the orientation of boundaries. These points are discussed in more detail in [16, 18].

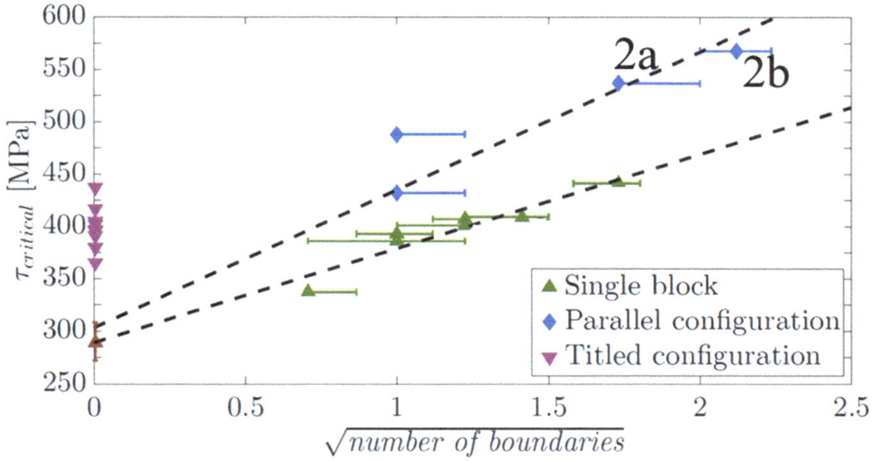

Fig. 4.4 A plot of τ_{CRSS} as a function of the square root of the number of boundaries. The *blue and purple data points* originate from multi-block samples which have there boundaries, respectively, parallel and tilted to the loading direction. The *green data points* represent single block specimens with sub-block boundaries parallel to the loading direction. The data points marked with numbers are from the samples which are shown in Fig. 4.2. Data partially reproduced from [16]

4.5 Conclusion

Uniaxial micro-tensile testing is conducted on lath martensite specimens with two configurations: block/sub-block boundaries roughly (1) parallel to the loading direction and (2) titled at approximately 45° to the loading direction. In the parallel configuration, both the block boundary and sub-block boundary function as a barrier for dislocation motion and fracture propagation, thereby strengthening the specimens. It was observed that the fracture surface cannot propagate across block boundaries, whereas sub-block boundaries are not capable of stopping the fracture surface. The critical resolved shear stress, τ_{CRSS}, shows a Hall–Petch-type relationship with the number of block boundaries and sub-block boundaries. Interestingly, when the boundaries are tilted to the loading direction, however, boundary sliding is an important new deformation mechanism, which seems to be activated easier than crystallographic slip.

References

1. Takaki, S., Ngo-huynh, K., Nakada, N., Tuchiyama, T.: ISIJ Int. **52**, 710–716 (2012)
2. Krauss, G.: Mater. Sci. Eng. A **273–275**, 40–57 (1999)
3. Ohmura, T., Hara, T., Tsuzaki, K.: Scr. Mater. **49**, 1157–1162 (2003)
4. Morito, S., Yoshida, H., Maki, T., Huang, X.: Mater. Sci. Eng. A **438–440**, 237–240 (2006)
5. Shibata, A., Nagoshi, T., Sone, M., Morito, S., Higo, Y.: Mater. Sci. Eng. A **527**, 7538–7544 (2010)
6. Zhang, C., Wang, Q., Ren, J., Li, R., Wang, M., Zhang, F., Sun, K.: Mater. Sci. Eng. A **534**, 339–346 (2012)
7. Ohmura, T., Hara, T., Tsuzaki, K.: J. Mater. Res. **18**, 1465–1470 (2003)
8. Shibata, A., Nagoshi, T., Sone, M., Morito, S., Higo, Y.: J. Alloy Compd. **577**, S555–S558 (2013)
9. Ghassemi-Armaki, H., Chen, P., Bhat, S., Sadagopan, S., Kumar, S., Bower, A.: Acta Mater. **61**, 3640–3652 (2013)
10. Ghassemi-Armaki, H., Maass, R., Bhat, S.P., Sriram, S., Greer, J.R., Kumar, K.S.: Acta Mater. **62**, 197–211 (2014)
11. Mine, Y., Hirashita, K., Takashima, H., Matsuda, M., Takashima, K.: Mater. Sci. Eng. A **560**, 535–544 (2013)
12. Ghadbeigi, H., Pinna, C., Celotto, S., Yates, J.R.: Mater. Sci. Eng. A **527**, 5026–5032 (2010)
13. Michiuchi, M., Nambu, S., Ishimoto, Y., Inoue, J., Koseki, T.: Acta Mater. **57**, 5283–5291 (2009)
14. Nambu, S., Michiuchi, M., Ishimoto, Y., Asakura, K., Inoue, J., Koseki, T.: Scr. Mater. **60**, 221–224 (2009)
15. Bergers, L.J., Hoefnagels, J.P.M., Geers, M.G.D.: J. Phys. D Appl. Phys. **47** (2014)
16. Du, C., Hoefnagels, J.P.M., Vaes, R., Geers, M.G.D.: Scr. Mater. **116**, 117–121 (2016)
17. Du, C., Hoefnagels, J.P.M., Bergers, L.J., Geers, M.G.D.: (submitted for publication) (2016)
18. Du, C., Hoefnagels, J.P.M., Vaes, R., Geers, M.G.D.: (submitted for publication) (2016)

Chapter 5
Evaluating Indent Pile-Up with Gold Films on Non-Plastically Deforming Substrates

MariAnne Sullivan and Bart Prorok

Abstract This work focused on ascertaining the effect of pile-up during indentation of thin films on substrates. Conventional understanding has postulated that differences in contact area resulting from pile-up or sink-in significantly alter the extraction of material properties. In this work, the specific case of pile-up with compliant, plastically deforming films on stiff, non-plastically deforming substrates was studied. Several literature methods to assess pile-up were leveraged, and a new technique was developed and validated to quantify projected pile-up. Indentation testing was performed on gold films of multiple thicknesses on several ceramic-based substrates. The results indicated that the degree of pile-up was solely a function of indent depth into the film. Pile-up was not influenced by film thickness or substrate elastic modulus. In other words, the pile-up development was insensitive to the presence of the substrate and how it contributes to the composite's elastic properties. In such case, if the elastic response of the film/substrate composite was independent of the degree of pile-up, then elastic data acquired from unloading did not require a contact area correction. The findings are confirmed using the Zhou–Prorok model for extracting film elastic properties for both gold and platinum films.

Keywords Thin films • Nanoindentation • Non-plastic • Pile-up • Sink-in

5.1 Introduction

Nanoindentation is a valuable tool when measuring elastic modulus and hardness of a material. With applications such as electronics becoming smaller, nanoindentation is even more important as a testing technique, particularly for thin films on substrates. Extracting exact material properties from these applications will allow for better determination of the materials before use, preventing what could be catastrophic failure in the sense of a product like a computer or cell phone. Many have utilized nanoindentation, and others are perfecting the method in the cases where there may be interference from the substrate. The first to allow for extraction of material properties were Oliver and Pharr [1, 2]. Modifications later came from Gao [3] and Doerner and Nix [4], with every iteration improving extraction of the film properties. Most recently, Zhou and Prorok made modifications to the idea from Doerner and Nix in order to create a weighting factor for both the film and substrate. This equation can be seen in Eq. (5.1):

$$\frac{1}{E} = \frac{1}{E_f}(1 - \Phi_s)\left(\frac{E_f}{E_s}\right)^{0.1} + \frac{1}{E_s}(\Phi_f) \tag{5.1}$$

where $\Phi_f = e^{-\alpha_f\left(\frac{t}{h}\right)}$ and $\Phi_s = e^{-\alpha_s\left(\frac{t}{h}\right)}$. With this improved model, solely material properties from the film and substrate are used, such as the elastic moduli of the film and substrate, E_f and E_s, and Poisson's ratio of film and substrate, α_f and α_s. From these parameters and the raw nanoindentation data, the correct film elastic modulus can be discerned [5, 6]. A major concern of researchers in the field of nanoscale mechanics has been pile-up and sink-in that may occur during nanoindentation of materials. It is particularly evident when indenting films on substrates. A stiff film on a compliant substrate can result in sink-in, whereas a complaint film on a stiff substrate results in pile-up. These can be seen in Fig. 5.1.

Others have investigated if this pile-up and sink-in affect the results of the indentation mechanical property information, as the projected area is typically used in Oliver–Pharr methods [1]. The equation comes from the slope of the unloading curve, seen in Eq. (5.2):

M. Sullivan • B. Prorok (✉)
Department of Mechanical Engineering, 275 Wilmore Laboratories, Auburn University, Auburn, AL 36849, USA
e-mail: prorok@auburn.edu

L. Starman et al. (eds.), *Micro and Nanomechanics*, Conference Proceedings of the Society for Experimental Mechanics Series,
DOI 10.1007/978-3-319-42228-2_5

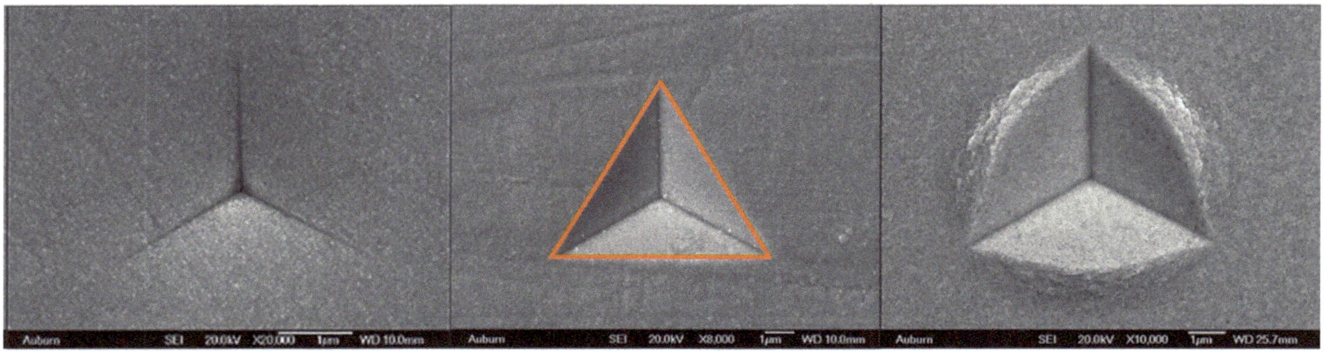

Fig. 5.1 SEM images of examples of sink-in, a normal Berkovich indent, and pile-up

$$S = \frac{dP}{dh} = \beta \frac{2}{\sqrt{\pi}} E_r \sqrt{A_c} \tag{5.2}$$

In this equation, S is the slope of the unloading curve, or change in load P over displacement h, β is a geometric term for the shape of the Berkovich indenter tip, E_r is reduced elastic modulus, and A_c is contact area. This area term is what other researchers are changing based on differences in the growth of pile-up or sink-in. Pile-up was examined first, as it is more easily determined from the projected area images with scanning electron microscope (SEM) images. Researchers such as Kese [7] and others [8] have attempted to characterize the pile-up through geometric measurements, deemed the semi-ellipse method, or predictions from models that are not empirical. However, this area term may not be necessary to change when utilizing the Zhou–Prorok model. Each indent was imaged in this work and measured in order to compare differences in pile-up based on film/substrate combinations of compliant films with non-plastically deforming substrates. The results will determine how the pile-up affects extraction of elastic properties.

5.2 Experimental Procedure

Samples were created using a Denton sputter coater with DC power. Films of gold and platinum were chosen due to their inert qualities along with ease of sputtering. Sputter time was controlled in order to obtain varying thicknesses of the films. Silicon was chosen as an initial substrate due to its use in previous work [5, 6] and well-studied nanoindentation. Other substrates tested were ceramics such as magnesium oxide (MgO), and aluminum oxide (Al_2O_3), and aluminum nitride (AlN). These were carefully chosen with varying elastic moduli yet all non-plastically deforming, removing plastic concerns during these particular indents. Following film deposition, nanoindentation was completed with an MTS Nanoindenter XP with displacement-controlled testing. An average of 25 indents in an array was taken from each sample. Based on the film thickness, the ratio of film thickness to indentation depth could also be compared. After nanoindentation tests, the indents were imaged with a JEOL 7000 F SEM. Additionally, cross-sectional cuts were created with a TESCAN LYRA FIB in order to view the same indents from a different perspective.

5.3 Results and Discussion

The first result from studying nanoindentation of indents from an aerial view showed a unique growth of pile-up with increasing indent depth. While this was expected, what was noteworthy was the fact that the Zhou–Prorok model was unaffected by pile-up. Most other researchers had been focused on measuring pile-up in order to later subtract it out from the elastic properties garnered from nanoindentation tests. Here, however, pile-up is very apparent on a series of indents of a gold film on silicon substrate. These are seen in Fig. 5.2. This is a single sample of 625 nm gold film on a silicon substrate, and it is evident the substrate is penetrating through around the 700 nm indent depth. It is interesting to follow the progression of the pile-up, the substrate punching through, and the pile-up scaling with indent depth.

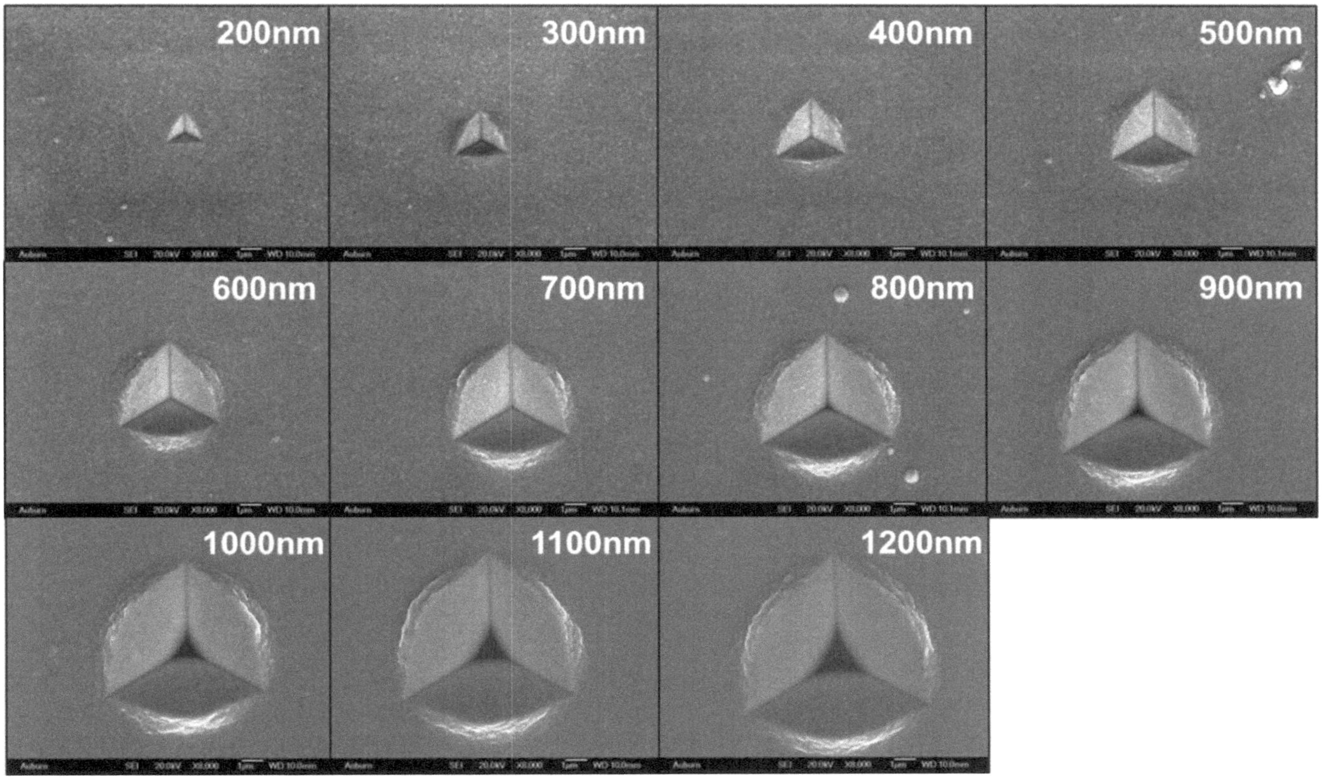

Fig. 5.2 Pile-up growth with indent progression of 625 nm gold film on silicon, SEM images taken for each indent depth noted

Utilizing Eq. (5.1), the Zhou–Prorok model is shown for different thicknesses. One sample for 625 nm is of the indent progression shown in the previous figure, and the others are thinner or thicker. Figure 5.3 demonstrates the Zhou–Prorok model and the extraction of the elastic modulus, shown as open symbols, for each film thickness of gold on silicon. It is apparent that the elastic modulus being extracted will change when there is a different film thickness, but the Zhou–Prorok equation can handle this because it considers a normalized thickness (h/t). However, pile-up occurs in every case as seen in the SEM images.

Further analysis of the substrate was necessary through FIB. An SEM aerial image of this is shown in Fig. 5.4a, and the cross section after an FIB cut is shown in Fig. 5.4b. The reason for this imaging shows the non-plastically deforming substrate of silicon with a gold film piling up. This confirms that there is minimal plastic deformation, making the calculations for this case much less complicated. In the future, however, these types of substrates that plastically deform, will be studied in the same manner. The cross-sectional view gives more information into how the film and substrate are acting together when indented.

The pile-up was to be investigated further with a variety of ceramic substrates besides solely silicon. The information from nanoindentation along with model predictions are seen in Fig. 5.5. The Zhou–Prorok model is still correctly extracting film elastic modulus, even with a range of substrate elastic moduli. Magnesia, alumina, and aluminum nitride are compared to the previous silicon in this figure with the same gold film thickness of 360 nm. These substrates are listed in order of their elastic moduli, and the model accurately predicts these differences, still extracting a constant film modulus for gold at every indent depth, seen with open symbols.

With the nanoindented samples at each indent depth, the pile-up was then considered for these ceramic substrates with gold films. The dotted lines are identical, and surprisingly, even with the changes in elastic moduli of the substrates, the pile-up is also identical between all four ceramic substrates seen in Fig. 5.6.

As this may not be as apparent, a method of measuring the pile-up with an ImageJ techniques, previously described [9, 10], a numerical area can be gathered from the pile-up seen in all of these cases. Figure 5.7 shows the four ceramic substrates and the measured amount of projected pile-up at each indent depth. They all align up to the film thickness, and even past penetrating into the substrate, they also appear to align.

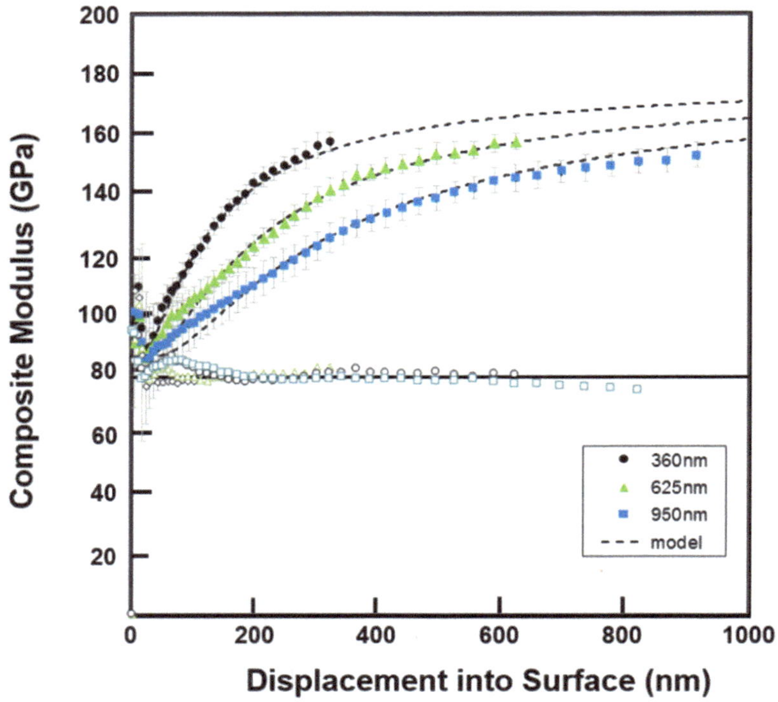

Fig. 5.3 Zhou–Prorok model for varying thicknesses of gold on silicon substrates, with extracted film modulus noted with *open symbols*

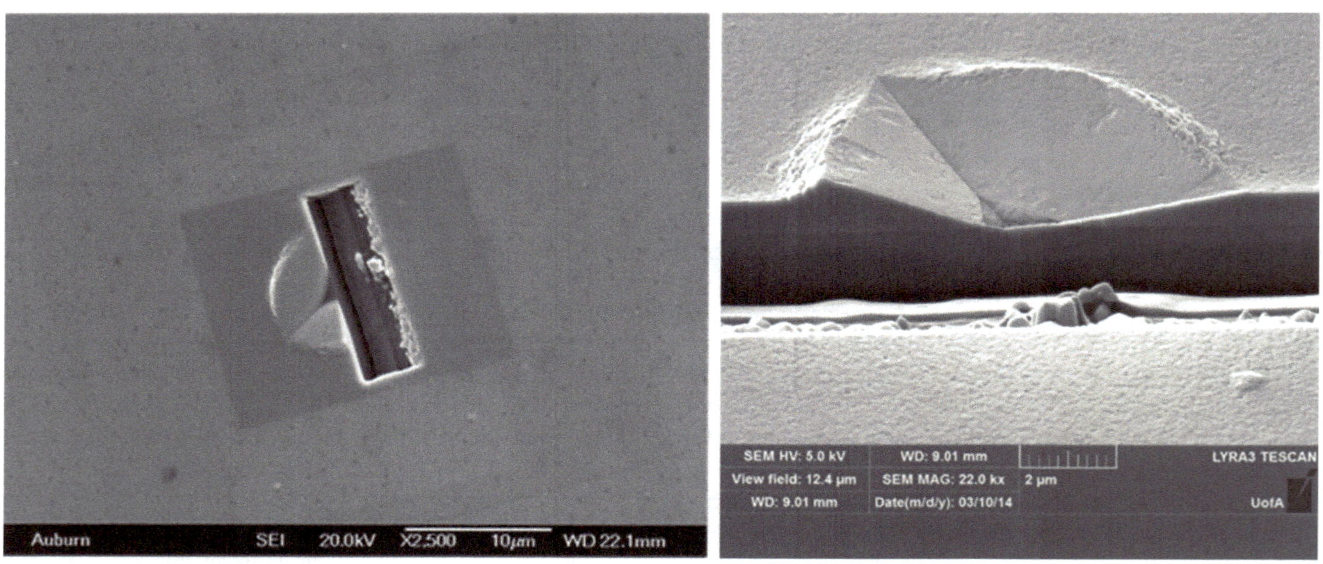

Fig. 5.4 SEM images taken after FIB cuts from above (**a**) and from an angle (**b**), with the film and the substrate visible

There are similar tendencies for platinum films, as well. These are all true for the particular case of compliant films on ceramic substrates. From the FIB image, along with images from SEM that show the substrate punching through, it is seen that the substrate is not plastically deforming. Future work will determine which film properties are controlling this pile-up, as there is less in platinum. Additionally, plastically deformed substrates will also be studied in order to decouple the elastic and plastic contributions to pile-up and sink-in during nanoindentation.

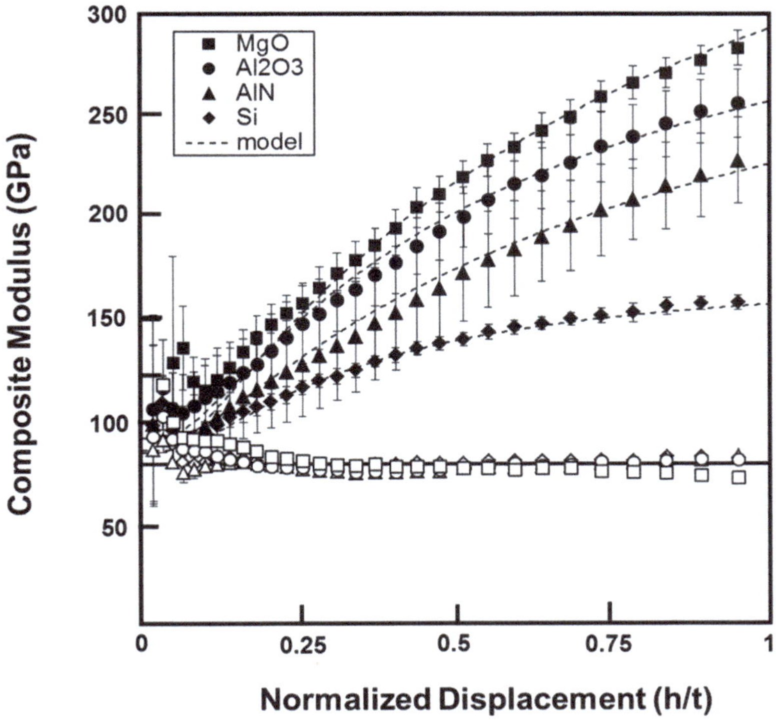

Fig. 5.5 Nanoindentation data compared with the Zhou–Prorok model for gold film on ceramic substrates denoted with *closed symbols*; *open symbols* represent the extracted film modulus from the model

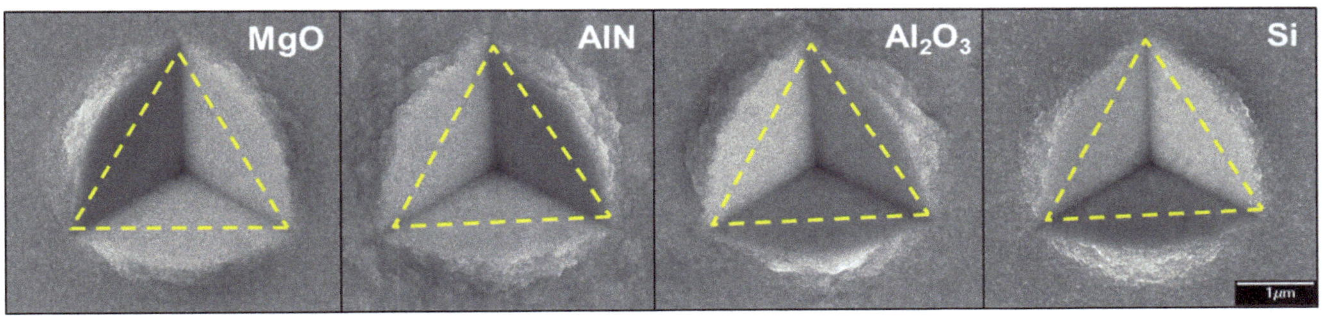

Fig. 5.6 SEM images of noted ceramic substrates with gold films compared to Berkovich contact area marked with *dotted lines*

5.4 Conclusions

Overall, the pile-up occurring from compliant films on non-plastically deforming substrates in these cases showed there was no effect on elastic properties garnered from nanoindentation. With the ceramic substrates, it was clear when the indenter was puncturing through to the substrate, and the Zhou–Prorok model was able to extract the elastic modulus up to the film thickness, and tends to hold true even beyond this point. With this newly collected information, nanoindentation of these types of samples, compliant films on non-plastically deforming substrates, no further manipulation of the contact area is needed with the Zhou–Prorok method. It was concluded that no matter the film thickness or substrate elastic modulus for the ceramic substrate samples, the pile-up was not affected. There is a similar trend on gold and platinum, but at varying degrees. This will lead to future work on plastically deforming substrates in order to discern the difference in elastic and plastic tendencies during nanoindentation and what other material properties may possibly be affected.

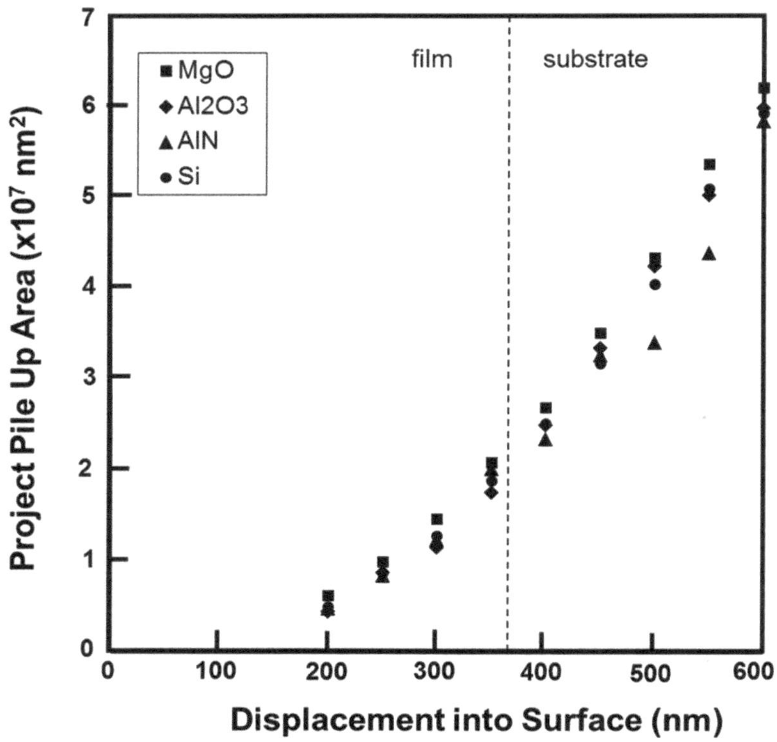

Fig. 5.7 Measurement of pile-up for displacement into surface for each ceramic substrate

References

1. Oliver, W.C., Pharr, G.M.: An improved technique for determining hardness and elastic modulus using load and displacement sensing indentation experiments. J. Mater. Res. **7**, 1564–1583 (1992)
2. Oliver, W.C., Pharr, G.M.: Measurement of hardness and elastic modulus by instrumented indentation: advances in understanding and refinements to methodology. J. Mater. Res. **19**(1), 3–20 (2004)
3. Gao, H., Chiu, C.-H., Lee, J.: Elastic contact versus indentation modeling of multi-layered materials. Int. J. Solids Struct. **29**, 2471 (1992)
4. Doerner, M.F., Nix, W.D.: A method for interpreting the data from depth-sensing indentation instruments. J. Mater. Res. **1**, 601–609 (1986)
5. Zhou, B., Prorok, B.: A discontinuous elastic interface transfer model of thin film nanoindentation. Exp. Mech. **50**, 793–801 (2010)
6. Zhou, B., Prorok, B.C.: A new paradigm in thin film nanoindentation. J. Mater. Res. **25**, 1671–1678 (2010)
7. Kese, K., Li, Z.C.: Semi-ellipse method for accounting for the pile-up contact area during nanoindentation with the Berkovich indenter. Scr. Mater. **55**, 699–702 (2006)
8. Bolshakov, A., Pharr, G.M.: Influences of pileup on the measurement of mechanical properties by load and depth sensing indentation techniques. J. Mater. Res. **13**(04), 1049–1058 (1998)
9. Sullivan, M., Prorok, B.C.: Evaluating indent pile-up with metallic films on ceramic-like substrate. J. Mater. Res. **30**(13), 2046–2054 (2015)
10. Sullivan, M., Prorok, B.C.: Newly discovered pile up effects during nanoindentation. MEMS Nanotechnol. **7**, 27–31 (2015)

Chapter 6
Investigation of Size Effect Through In-Situ SEM Testing of Polystyrene Micropillars

Thimmappa Shetty Guruprasad, Shantanu Bhattacharya, and Sumit Basu

Abstract Size effects in micron sized pillars loaded in compression, fabricated out of metal single crystals with FIB milling has been reported by several authors. The size effect manifests in a significant amplification of the flow stress with decrease in diameter. Amorphous glassy polymers are widely used in micro-structures, such as in many micro-electro-mechanical system (MEMS) based devices in electronics and biological applications. But their mechanical behaviours at these length scales have not been adequately explored. In this work, we show that micropillars of an amorphous material, polystyrene (PS), also exhibit significant size effects under compression. The pillars are fabricated by methods that do not involve focused ion beam (FIB) milling and are therefore, free of ion damage. Micropillars of diameters ranging from 38 to 190 μm have been fabricated using a microfabrication technique and they are deformed in-situ in a FESEM under compression. The stress strain plots are obtained by fitting the load displacement behaviour using well calibrated constitutive model. Results suggest that the yield stress of PS intrinsically depend on the diameter.

Keywords Size effect • Polystyrene • Micro-pillars • SEM • Micro-compression

6.1 Introduction

Polymers are increasingly being used as functional/structural materials in MEMS devices. Electronics and opto-electronics are changing materials strongly towards polymers, including displays, photo voltaic devices, memory, and many more [9]. Polymers offers good biocompatibility, which is necessary for some of the biomedical applications in microfluidic systems. Also, the transparency of many polymers is an advantage if visual control is necessary. In addition to all this, it offers good mechanical properties.

With increasingly use of materials at micron scale, we come across many situations, where strength of the material depend on the dimensions of the sample. One of the striking example is demonstrated by Uchic et al. [11] on cylindrical Ni micro-pillars machined by FIB milling. The methodology of uniaxial compression to study mechanical behavior of micron-sized samples is used, and they reported that strength determined to be many times higher than bulk material. Subsequently, several authors have reported size effects in micron sized pillars loaded in compression, fabricated out of metal single crystals (see Uchic et al. [12] for a review). The size effect manifests in a significant amplification of the flow stress with decrease in diameter. Though the origins of this effect is a subject of intense debate, it is generally believed that a "dislocation starvation mechanism" operates where dislocations exit the surface of the crystal at a rate faster than they can be nucleated within its bulk [5].

Following this micro-compression approach, several authors have reported testing of metallic glass to study size effect. Similar to crystalline materials, they fabricated metallic glass micropillars by FIB milling. Metallic glasses are amorphous materials and dislocation based explanations on size effect does not apply. There is no consensus, how size effect is observed in these materials. Increase in yield stress with decrease in diameter from 100 nm to 3 μm has been reported by Wang et al. [13] for $Al_{88}Fe_7Gd_5$ metallic glass. Also, many authors have reported size effect in metallic glasses [2, 7, 8].

Polymers are important class of amorphous materials. But, studies on polymers at micron scale is not adequately explored. He et al. has conducted compression of PS-DVB particles of diameter ranging from 2.6 to 25.1 μm using flat punch nanoindentation and observed significant increase in compressive strength with decrease in particle size. Recently, size effect on FIB machined epoxy micro-pillars in compression for diameters ranging from 0.7 to 5 μm have been observed by Wang et al. [14]. They reported significant increase in yield and fracture strength of epoxy micropillars with decrease in diameter.

T.S. Guruprasad (✉) • S. Bhattacharya • S. Basu
Department of Mechanical Engineering, Indian Institute of Technology Kanpur, Kanpur 208016, Uttar Pradesh, India
e-mail: tgprasad@iitk.ac.in

L. Starman et al. (eds.), *Micro and Nanomechanics*, Conference Proceedings of the Society for Experimental Mechanics Series,
DOI 10.1007/978-3-319-42228-2_6

Polystyrene (PS) is one of the most widely used polymers. It has many potential applications in microfluidic systems, especially in optical microfluidic devices. The increased use of PS is mainly due to its high optical-transparency in the visible light spectrum, biocompatibility, inertness, rigidity, and its surface modifiability [3].

In this work, we fabricated PS micropillars using microfabrication technique, which do not involve FIB milling and are free of damage processing issues. We fabricated PS micropillars of diameters ranging from 38 to 190 μm, with aspect ratio of 2:1. The fabricated micro-pillars are then deformed in-situ in a FESEM under compression using custom made micro-compression machine. The stress strain plots for all the diameters are obtained by fitting the experimental load displacement behaviour with a constitutive model for amorphous polymers given by Boyce et al. [1]. The constitutive model is implemented in ABAQUS.

6.2 Constitutive Model

To model the micropillar experiments of PS, the rate dependent elastic-viscoplastic constitutive model given by Wu and van der Giessen [15] is used. The constitutive model is based on the formulation of Boyce et al. [1], but we use a modified version introduced by Wu and van der Giessen [15].

The constitutive model takes the form,

$$\overset{\nabla}{\boldsymbol{\sigma}} = \boldsymbol{\mathcal{L}}_e : \boldsymbol{D}_e, \tag{6.1}$$

$\boldsymbol{\mathcal{L}}_e$ is the fourth-order isotropic elastic tensor, $\boldsymbol{D}_e$ is the elastic rate of strain tensor and $\overset{\nabla}{\boldsymbol{\sigma}}$ is the Jaumann rate of the Cauchy stress given in terms of the spin tensor $\boldsymbol{W}$ as

$$\overset{\nabla}{\boldsymbol{\sigma}} = \dot{\boldsymbol{\sigma}} - \boldsymbol{W}\boldsymbol{\sigma} + \boldsymbol{\sigma}\boldsymbol{W}, \tag{6.2}$$

Assuming that the yield response is isotropic, the isochoric visco-plastic strain rate

$$\boldsymbol{D}_p = \frac{\dot{\gamma}_p}{\sqrt{2}\tau}\bar{\boldsymbol{\sigma}}', \tag{6.3}$$

where $\bar{\boldsymbol{\sigma}}'$ is the deviatoric stress tensor, $\dot{\gamma}_p$ is the equivalent plastic strain rate, $\dot{\gamma}_p = \sqrt{\boldsymbol{D}_p : \boldsymbol{D}_p}$ and τ is the equivalent shear stress, $\tau = \sqrt{\frac{1}{2}\bar{\boldsymbol{\sigma}}' : \bar{\boldsymbol{\sigma}}'}$.

Using the rate tangent formulation for this constitutive model due to Peirce et al. [10] and Wu and van der Giessen [16], the constitutive model is implemented using UMAT feature in ABAQUS v6.11. The methods of fitting the parameters in detail are available at [6].

6.3 Experiments and Numerical Simulations

6.3.1 Fabrication of Polystyrene Micropillars

First, PS beads are dissolved in toluene. PS beads are procured from Sigma–Aldrich, has a weight average molecular weight, $M_w = 380{,}000$ and number average molecular weight, $M_n = 170{,}000$. PS thin films are prepared by dip coating glass slides into the dissolved solution. PS thin films are allowed to dry for 48 h in vacuum desiccator. Here, the solution of 50 % w/v of PS beads in toluene are made and obtained a PS thin films of thickness in the range of 250–300 μm.

Polydimethylsiloxane (PDMS) mold for hot embossing process is made by casting PDMS solution on SU-8 pillars. PS micropillars are fabricated by now placing PS thin film over the PDMS mould, then the assembly is sandwiched between two glass slides. The glass slides are lightly clamped together. The whole assembly is kept in oven at 180 °C for 20 min. Then, the assembly is allow to cool down to room temperature. Finally, PS thin film is carefully removed from PDMS mold [4] to get PS micropillars. The complete fabrication steps are shown in Fig. 6.1.

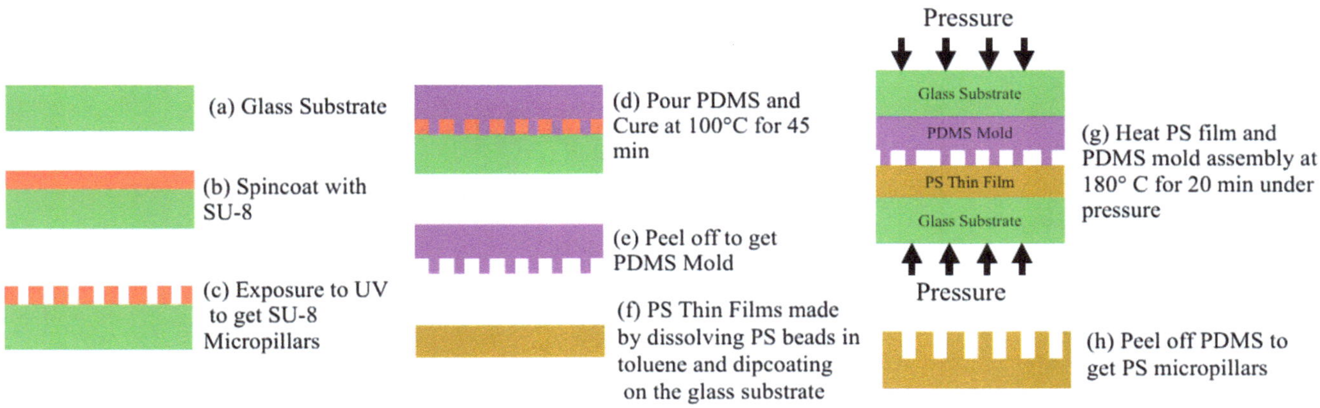

Fig. 6.1 Steps involved in PS micropillar fabrication

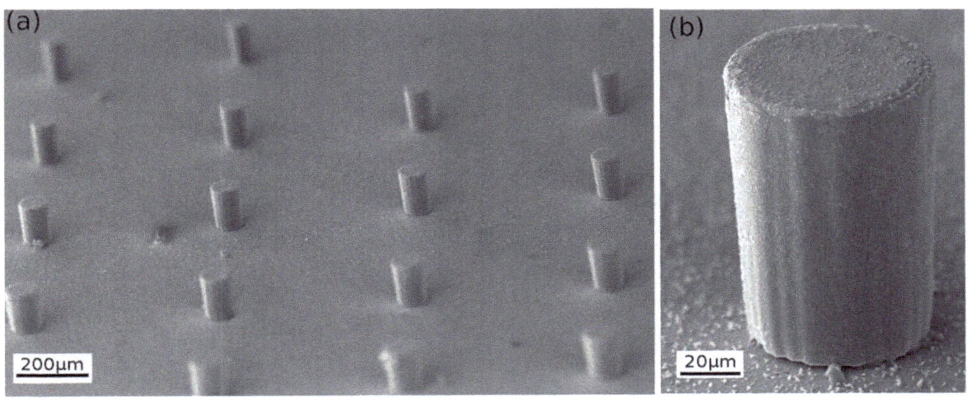

Fig. 6.2 (**a**) FESEM images of array of micropillars of diameter from 66 to 95 µm. (**b**) Single micropillar of diameter 66 µm

The Field Emission Scanning Electron Microscopic (FESEM) images of array of PS micropillars are shown in Fig. 6.2. We fabricated array of micro-pillars of diameters ranging from 38 to 190 µm, with a aspect ratio of 2:1. The fabricated micropillars sits on a 200 µm thick layer of PS.

6.3.2 Testing Procedure

The microcompression testing of micropillars are performed under the FESEM using a custom made microcompression device, the device is shown in Fig. 6.3. The microcompression device is a vacuum compatible and capable to do in-situ tests inside a SEM. The major components of the device are miniaturized load cell for force measurement, high resolution nanopositioner for displacement actuation and high resolution motorized five axis aligner for micro specimen alignment. The nanopositioner has a resolution in sub-nanometer range, load cell has a resolution of 133 µN and motorized aligner has a resolution of 30 nm. This device can test between a displacement rate of 50 nm/s to 13 mm/s and can be tested up to a maximum load of 2.5 N. A truncated steel nanoindenter with a flat top of diameter 150 and 500 µm was used for compressing the micro-pillars. Firstly, the sample is mounted on a motorized 5-axis aligner, then the fabricated micro-pillars are aligned with the electron beam using 5-axis aligner. The compression tests are then performed in the displacement controlled mode. The micro pillars of diameters ranging from 38 to 190 µm are tested at a strain rate of $4 \times 10^{-3}\,s^{-1}$. The nanopositioner is controlled by Labview. Also, the readings from load cell are synchronized with nanopositioner readings in Labview.

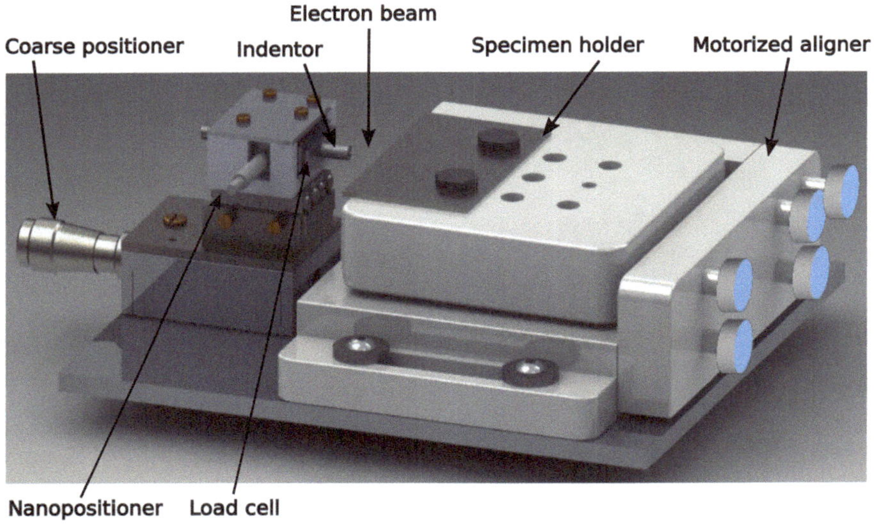

Fig. 6.3 A schematic of the custom made micro-compression device

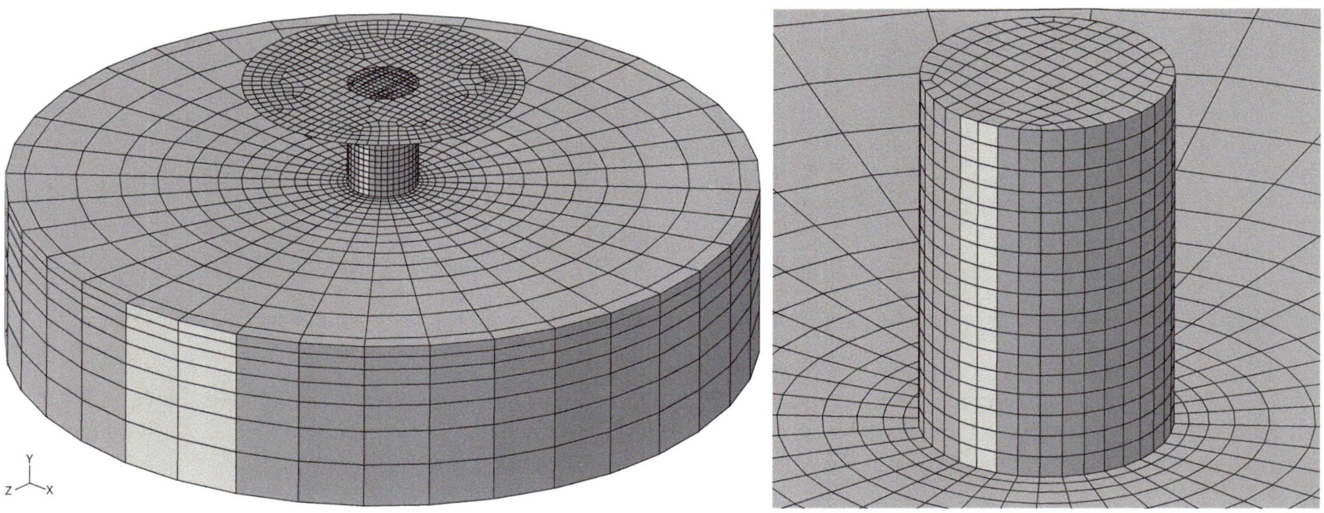

Fig. 6.4 Micropillar meshed model

6.3.3 Finite Element (FE) Simulations

Three dimensional FE simulations have been performed using the commercial FE software ABAQUS v6.11. Initially, PS micropillars of diameters ranging from 38 to 190 μm are modeled. Then material model for PS given in Sect. 6.2 is implemented in ABAQUS using UMAT feature. The finite element mesh of micropillar is shown in Fig. 6.4. In the formulation of boundary conditions, the bottom surface of film is not allowed to move in Y axis. Micropillars are then deformed in compression by applying a velocity in Y direction using a rigid flat indenting surface. For all the diameters of tested PS micropillars, their respective geometry is modeled and are simulated with same experimental conditions.

6.4 Results and Discussions

We fabricated PS micro-pillars of diameter ranging from 38 to 190 μm and are deformed in-situ in an FESEM under compression. The load displacement behaviour of micro-pillars for diameters 38 and 45 μm is shown in Fig. 6.5a. For each of the diameters, three specimens are tested and the error bar in load at yield point versus micropillar diameter is shown in

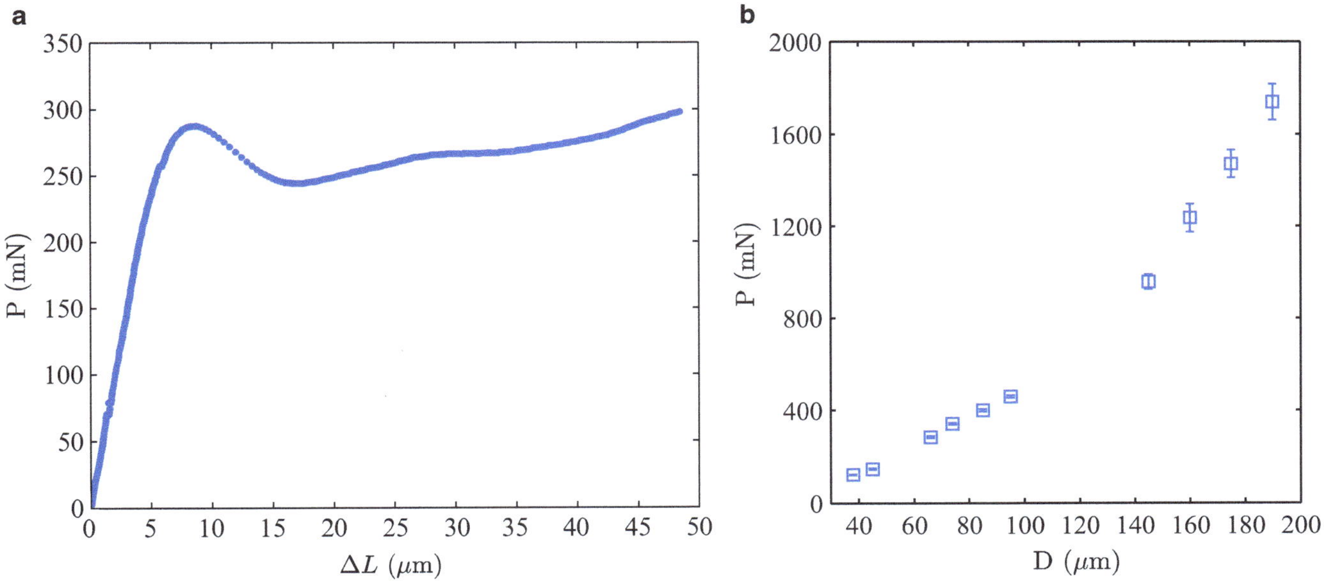

Fig. 6.5 (**a**) Load displacement (P$-\Delta L$) behavior of 66 μm micropillar diameter (**b**) Error bar in load at yield point versus pillar diameter (D)

Fig. 6.6 FESEM images of 66 μm micro pillar (**a**) before and (**b**) after 40 % compression

Fig. 6.5b. From Fig. 6.5a, we can see 66 μm diameter exhibit the classical features from a compression test on PS, namely, the linear elastic deformation, yield drop followed by rehardening response.

Figure 6.6 shows the FESEM images of 66 μm micro-pillar before and after compression. Though the sample does barrel towards the final stages of the experiment, it undergoes a homogeneous deformation for most part of it, nearly up to an engineering strain of 30 %. It is surprising given that the microcompression tests are somewhat different from tests generally conducted on bulk samples. The bottom face of the micropillar is on a thin film of PS instead of a hard surface and probably cannot expand freely. Also, the friction between the flat face of the indenter and the top face of the sample is not controlled.

Key parameters in the constitutive model are varied to obtain fits to the experimental load displacement responses using the ABAQUS, for all the diameters ranging from 38 to 190 μm. The parameters of constitutive model are noted for all the diameters. Two representative fits are shown in Fig. 6.7. We have fitted the initial parts of the load displacement responses very carefully, since we are interested in the yield stress and the later parts, especially for larger diameters showed complicated deformation response. The parameters governing the initial part of the response, are not affected by the parameters of the constitutive model that control the later part of the deformation.

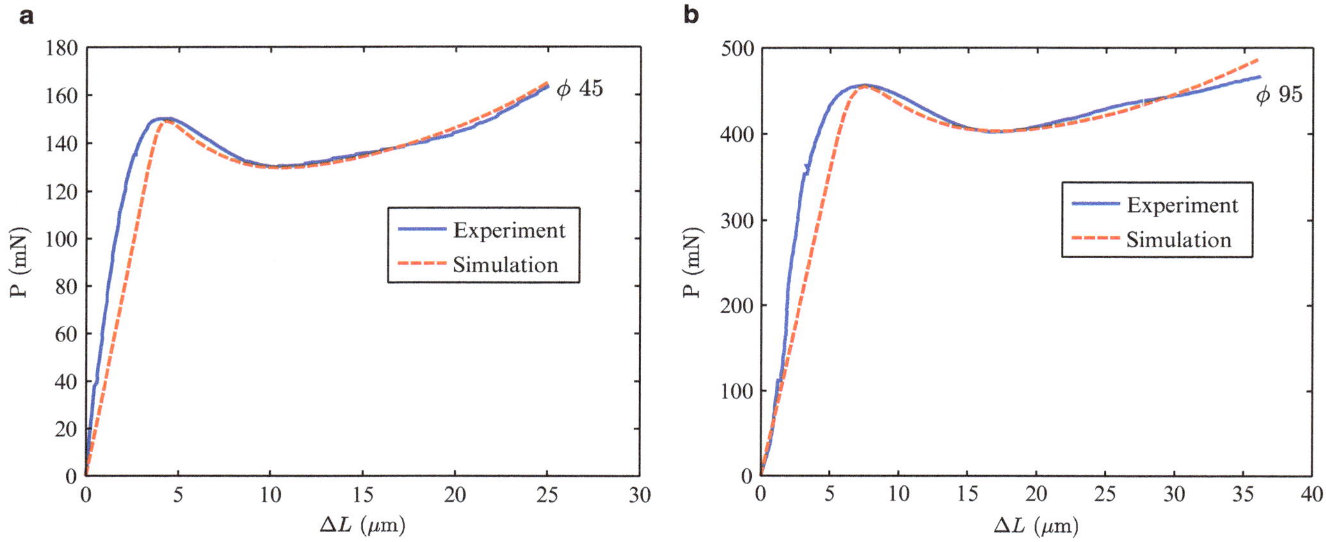

Fig. 6.7 Comparison of experimental and simulation load displacement curves for diameters (**a**) 45 μm and (**b**) 95 μm

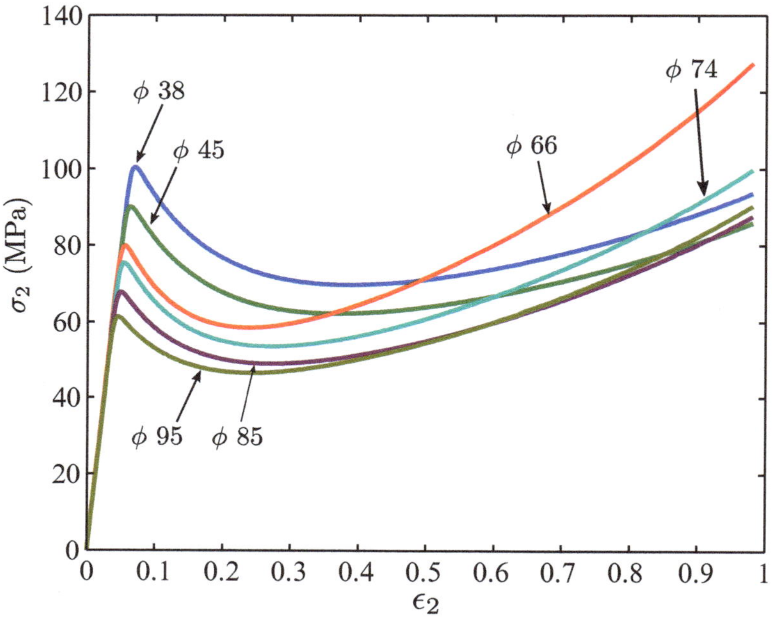

Fig. 6.8 Stress strain behaviours of micropillars of diameters ranging from 38 to 95 μm tested under uniaxial compression

The constitutive model accounts only for linear elastic behaviour while some amount of nonlinear elastic behaviour exhibited by micropillar during compression. For all the pillars, we have fitted secant modulus using E = 1600 MPa. In this process, both load at yield point and displacement at which it is attained are same for both experiment and simulation, which is shown in Fig. 6.7.

Stress strain plots for all the diameters are obtained by using the parameters pertaining to their respective diameters. The yield stress at the constant imposed strain rate of $4 \times 10^{-3}\,s^{-1}$, clearly increases with decrease in diameter which is shown in Fig. 6.8. The constitutive model parameter used to fit the initial response has to be increased steeply with decrease in diameter in order to fit the load displacement responses.

6.5 Conclusions

PS micropillars are fabricated by microfabrication technique and which do not involve FIB milling and are therefore, free of ion damage. PS micropillars of diameters ranging from 38 to 190 μm have been tested in-situ inside a FESEM under compression. The constitutive model parameters for PS have been varied using a ABAQUS UMAT on pillars of same dimensions to match experimental load displacement behaviour. Stress strain response from the load displacement fits for micropillar diameters reveals the existence of size dependent yield stress in PS.

References

1. Boyce, M.C., Parks, D.M., Argon, A.S.: Large inelastic deformation of glassy polymers. Part I: rate dependent constitutive model. Mech. Mater. **7**(1), 15–33 (1988)
2. Cheng, S., Wang, X.-L., Choo, H., Liaw, P.K.: Global melting of Zr57Ti5Ni8Cu20Al10 bulk metallic glass under microcompression. Appl. Phys. Lett. **91**(20):201917 (2007)
3. Fan, Y., Li, H., Yi, Y., Foulds, I.G.: PMMA to polystyrene bonding for polymer based microfluidic systems. Microsyst. Technol. **20**(1), 59–64 (2014)
4. Goral, V.N., Hsieh, Y.-C., Petzold, O.N., Faris, R.A., Yuen, P.K.: Hot embossing of plastic microfluidic devices using poly(dimethylsiloxane) molds. J. Micromech. Microeng. **21**(1), 017002 (2011)
5. Greer, J.R., Oliver, W.C., Nix, W.D.: Size dependence of mechanical properties of gold at the micron scale in the absence of strain gradients. Acta Mater. **53**(6), 1821–1830 (2005)
6. Kattekola, B., Desai, C.K., Parameswaran, V., Basu, S.: Critical evaluation of a constitutive model for glassy polycarbonate. Exp. Mech. **54**(3), 357–368 (2014)
7. Lai, Y.H., Lee, C.J., Cheng, Y.T., Chou, H.S., Chen, H.M., Du, X.H., Chang, C.I., Huang, J.C., Jian, S.R., Jang, J.S.C., Nieh, T.G.: Bulk and microscale compressive behavior of a Zr-based metallic glass. Scripta Mater. **58**(10), 890–893 (2008)
8. Lee, C.J., Huang, J.C., Nieh, T.G.: Sample size effect and microcompression of Mg65Cu25Gd10 metallic glass. Appl. Phys. Lett. **91**(16), 161913 (2007)
9. Liu, C.: Recent developments in polymer MEMS. Adv. Mater. **19**(22), 3783–3790 (2007)
10. Peirce, D., Shih, C.F., Needleman, A.: A tangent modulus method for rate dependent solids. Comput. Struct. **18**(5), 875–887 (1984)
11. Uchic, M.D., Dimiduk, D.M., Florando, J.N., Nix, W.D.: Sample dimensions influence strength and crystal plasticity. Science **305**(5686), 986–989 (2004)
12. Uchic, M.D. , Shade, P.A., Dimiduk, D.M.: Plasticity of micrometer-scale single crystals in compression. Ann. Rev. Mater. Res. **39**(1), 361–386 (2009)
13. Wang, C.-C., Ding, J., Cheng, Y.-Q., Wan, J.-C., Tian, L., Sun, J., Shan, Z.-W., Li, J., Ma, E.: Sample size matters for Al88Fe7Gd5 metallic glass: smaller is stronger. Acta Mater. **60**(13–14), 5370–5379 (2012)
14. Wang, S., Yang, Y., Zhou, L.M., Mai, Y.-W.: Size effect in microcompression of epoxy micropillars. J. Mater. Sci. **47**(16), 6047–6055 (2012)
15. Wu, P.D., Vander Giessen, E.: On the improved network model for rubber elasticity and their applications to orientation hardening in glassy polymers. J. Mech. Phys. Solids **41**(3), 427–456 (1993)
16. Wu, P.D., Vander Giessen, E.: Computational aspects of localized deformations in amorphous glassy polymers. Eur. J. Mech. A/Solids **15**(5), 799–823 (1996)

Chapter 7
Temperature and Thickness Dependent Mechanical Properties of Ti/Ni Multilayer Thin Films

Zhou Yang and Junlan Wang

Abstract Strong temperature and thickness dependent mechanical properties of Ti/Ni multilayer thin films have been observed with layer thickness from 200 nm to 6 nm and annealing temperature from room temperature to 500 °C. The as-deposited case follows the traditional trend of dislocation mediated-strengthening to grain boundary mediated-softening with decreasing layer thickness. Initial thermal strengthening of multilayers is achieved by annealing induced grain boundary relaxation. This strengthening is found to increase with decreasing layer thickness and increasing annealing temperature. Further strengthening could be achieved due to solid solution of diffused atoms and Ti-Ni intermetallic precipitates for multilayers with thin layer, while obvious softening has been observed for multilayers with thick layer due to recrystallization and grain growth.

Keywords Ti/Ni • Multilayer thin film • Annealing temperature • Layer thickness • Hardness

7.1 Introduction

Thermal/mechanical stability of nanoscale metallic multilayer thin films have attracted more and more attention due to the advanced properties and wide applications of multilayered material systems [1–6]. An obvious individual layer thickness dependent hardness behavior is reported in extensive studies [7–11]. The multilayer strength increases dramatically with decreasing layer thickness due to dislocation mediated mechanisms [12–16], and decreases with further reduced layer thickness due to grain boundary mediated mechanisms [17, 18]. In addition, thermal annealing has found to have an important influence on the deformation behavior and strengthening mechanism in metallic multilayers. Depending on the annealing temperature, various thermal processes could take place in multilayer systems, such as interdiffusion and intermixing along interface, recrystallization and grain growth, phase transformation, and intermetallic formation. The thermal/mechanical stability at elevated temperatures could be significantly influenced if the materials were exposed to thermal environment. Previous works observed decreased material strength in various multilayer thin films after a relatively high temperature annealing [19–21], due to heat-induced microstructure modifications such as in-plane grain coarsening of Ni/Ru multilayer films annealed at 450 °C [20] and spheroidization in Cu/Ag multilayer films annealed after 150 °C [19]. In addition, significant morphological instabilities including complete disintegration of layered structure [22–24] were observed in different multilayer thin films. However, excellent thermal stability (up to 700 °C) was observed for Ni/Ru multilayer with thick layer (>35 nm) [25]. The diffusion based instability mechanism was recently studied for typical HCP-based nanoscale multilayer system [26]. Considering the strong effect from temperature and thickness on aforementioned thermal/mechanical properties as well as the possible alternation of functional properties, systematical studies on temperature and thickness dependent thermal/mechanical stability are needed.

In current work, Ti/Ni multilayer thin films were selected as a model system with a wide layer thickness range from 200 nm down to 6 nm, and the applied annealing temperature range up to 500 °C. The current work investigates the detailed temperature and thickness dependent mechanical properties of nanoscale multilayer thin films, and the corresponding strengthening mechanisms. The results not only shed light on the scientific understanding of the thermal/mechanical stability in metallic multilayers, but also provide guidance to control and optimize this new engineering material.

Z. Yang • J. Wang (✉)
Department of Mechanical Engineering, University of Washington, Box 352600, Seattle, WA 98195, USA
e-mail: junlan@u.washington.edu

L. Starman et al. (eds.), *Micro and Nanomechanics*, Conference Proceedings of the Society for Experimental Mechanics Series,
DOI 10.1007/978-3-319-42228-2_7

7.2 Experimental Details

Magnetron sputtering system (Orion-5-UHV, AJA International, MA) is used to deposit Ti/Ni multilayer thin films on single crystal (100) Si substrates, with the background pressure around 1×10^{-7} mbar and deposition pressure around 1×10^{-3} mbar. During deposition, both the DC powers of Ti (99.995 %) and Ni (99.999 %) targets were 100 W and Ar gas flow rate 10 sccm. The sputtering time was controlled for both Ti and Ni in order to obtain the desired layer thickness. The same individual layer thickness is maintained in each sample. Samples are divided into three groups based on the range of layer thickness, λ: thick-layer cases with λ of 200 and 100 nm, intermediate-layer cases with λ of 50 and 25 nm, and thin-layer cases with $\lambda \leq 16$ nm. The nominal total film thickness is 4 μm in the thick- and intermediate-layer cases and 2 μm in the thin-layer case. A set of the as-deposited Ti/Ni multilayer films were annealed in Ar atmosphere for 2 h with varying temperatures from 100 to 500 °C.

An Ubi1 nanomechanical test instrument (Hysitron, Inc., MN) with a cube corner indenter tip was utilized to obtain the hardness for both as-deposited and annealed multilayer films. The indentation followed a trapezoidal loading function including a 10 s loading, 5 s hold and 10 s unloading. A standard Oliver and Pharr method [27] was used to obtain the average hardness from a minimum of 25 indents. For microstructural characterization, X-ray diffraction (XRD) with Cu Kα line was used to detect the phase information, and scanning electron microscopy (SEM) was used to examine the surface and cross-section morphological evolution before and after annealing. For the thin layer cases, transmission electron microscopy (TEM) was used to obtain the cross-section morphology and diffraction pattern for selected samples.

7.3 Results and Discussions

7.3.1 *Microstructure Characterization*

Figure 7.1a shows the XRD spectra of as-deposited Ti/Ni multilayer thin films with individual layer thickness from 200 nm to 6 nm. In the thick-layer cases, the XRD pattern shows a FCC structure for Ni with strong peaks at 44.3° corresponding to the reflection from Ni (111), 51.7° from Ni (200), and 76.2° from Ni (220), while a HCP structure for Ti with a strong peak at 38.6° corresponding to the reflection from Ti (002). In the intermediate-layer cases, the intensities of Ti (002) peak and Ni (200) become smaller and smaller, and Ni (220) peak disappears. In the thin-layer cases, Ti peaks become weaker with decreased layer thickness, and no longer distinguished with layer thickness less than 12 nm. Only Ni (111) peak maintains significant and the other two Ni peaks Ni (200) and Ni (220) are not visible. The large intensity difference between Ni and Ti peaks indicate that the Ni layer is more crystalline and textured than Ti layer. Peak broadening with decreasing layer thickness indicates finer grains for samples with thinner layer. Grain growth would be suppressed in thin layer case and fine grains would dominate.

Annealing induced microstructure modification can be revealed by XRD patterns. For samples with $\lambda = 100$ nm, as shown in Fig. 7.1b, similar XRD spectra were obtained after annealing up to 200 °C, while the Ti peak almost disappeared after 300 °C annealing. A couple of Ti-Ni intermetallic phases were detected after 500 °C annealing. However, a different phenomenon is displayed in the samples with $\lambda = 8$ nm as shown in Fig. 7.1c. Ni (111) is the only significant peak shown

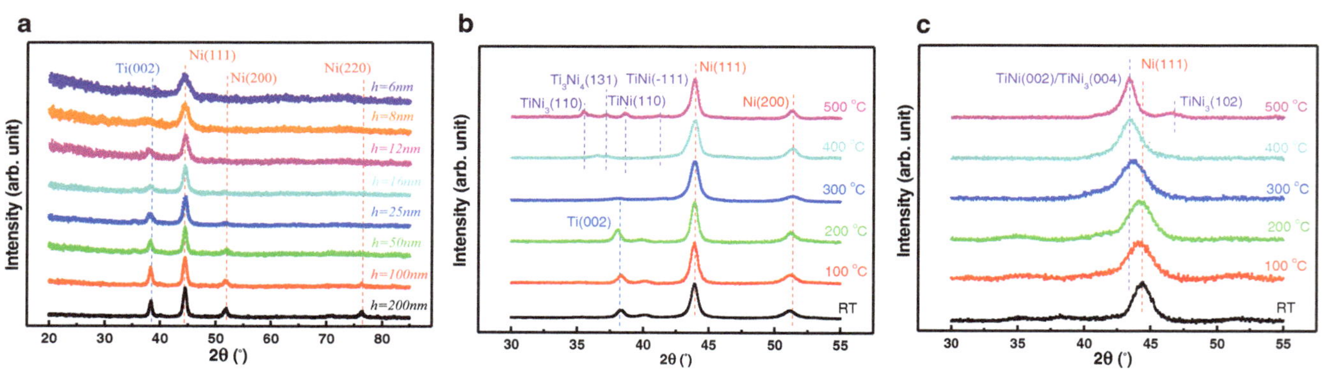

Fig. 7.1 XRD spectra of (**a**) as-deposited and annealed Ti/Ni multilayers with (**b**) $\lambda = 100$ nm and (**c**) $\lambda = 8$ nm

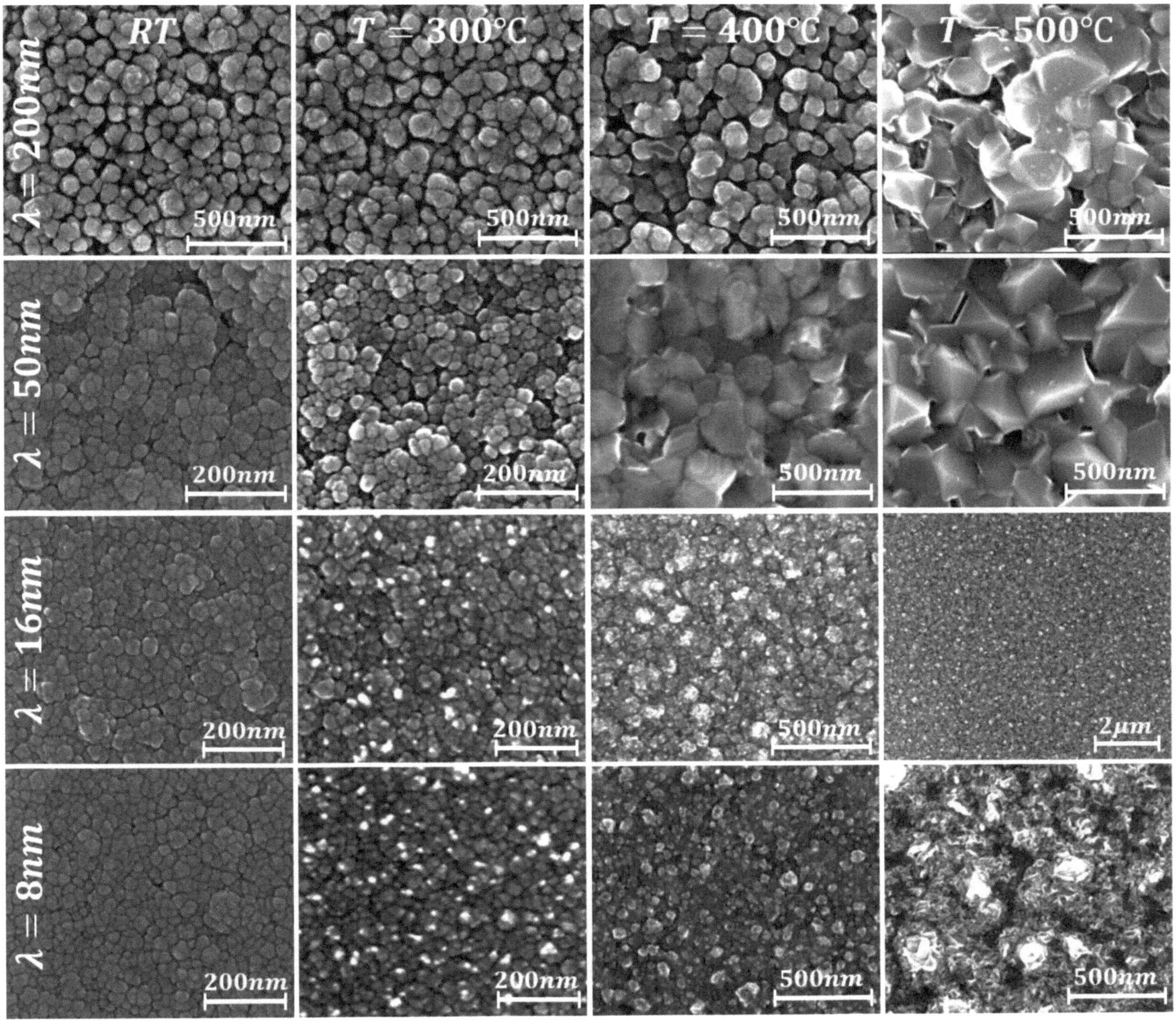

Fig. 7.2 SEM images of surface morphology evolution of Ti/Ni multilayers with λ = 200 nm, λ = 50 nm, λ = 16 nm and λ = 8 nm

in the as-deposited sample. An obvious peak shifting is observed after 300 °C annealing, and this broad peak is likely a combination of the intermetallic peak and Ni (111) peak. After high temperature annealing, only Ti-Ni intermetallic peaks were detected, indicating the possible disintegration of the layers and the formation of fully intermixed Ti-Ni alloy.

Figure 7.2 shows annealing induced surface morphological evolution for Ti/Ni multilayer thin films with layer thickness 200 nm, 50 nm, 16 nm and 8 nm. For as-deposited cases, homogeneous surface structure with uniform grains was shown, and no significant change after low temperature annealing with $T < 300$ °C [28]. After 300 °C, obvious diffusion was observed in the thin layer cases, such as $\lambda = 16$ nm and $\lambda = 8$ nm cases, while still no measureable change in $\lambda = 50$ nm and $\lambda = 200$ nm cases. After high temperature annealing with $T > 300$ °C, a very different surface morphology is formed in the thin layer cases, probably corresponding to the Ti-Ni alloys, while recrystallization and obvious grain growth in thick- and intermediate-layer cases. The surface shows initial recrystallization after 400 °C and fully recrystallized grain structures after 500 °C in the $\lambda = 50$ nm case, while recrystallization upon 500 °C in the $\lambda = 200$ nm case.

Figure 7.3 shows the cross-section morphology for both as-deposited and annealed Ti/Ni multilayer thin films with different layer thickness. For as-deposited cases, layered structure can be clearly observed, as shown in TEM image in Fig. 7.3a for $\lambda = 12$ nm case, and SEM images in Fig. 7.3d for $\lambda = 200$ nm case and Fig. 7.3h for $\lambda = 25$ nm case. Figure 7.3b shows the cross-section morphology for 200 °C annealed $\lambda = 16$ nm sample, and distinct interface can be observed in the

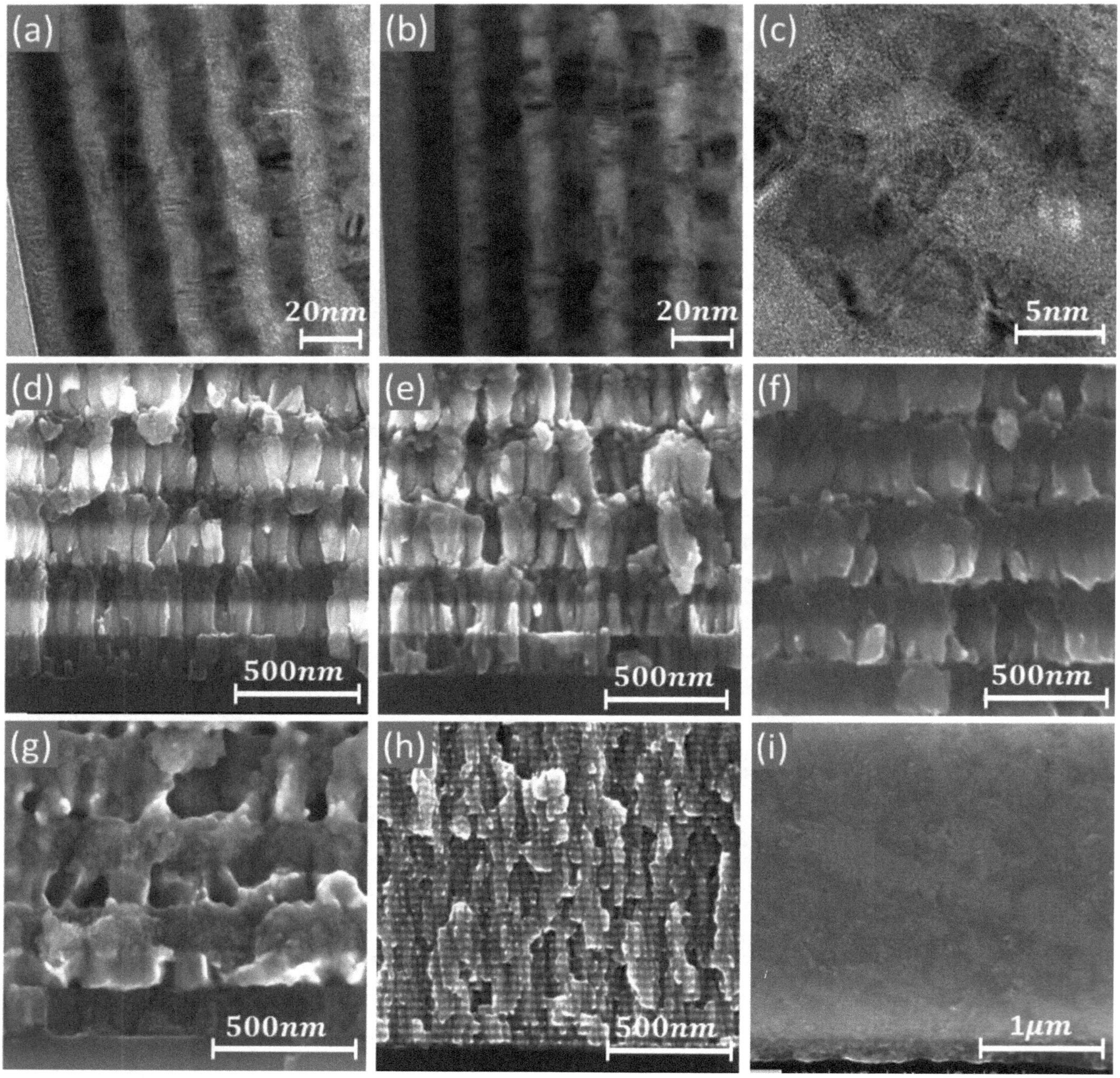

Fig. 7.3 Cross-section morphology of Ti/Ni multilayers in λ = 8 nm case (**a**) as-deposited, (**b**) 200 °C annealed, (**c**) high resolution TEM image for (**b**); λ = 200 nm case (**d**) as-deposited, (**e**) 200 °C annealed (**f**) 400 °C annealed, (**g**) 500 °C annealed; and λ = 25 nm case (**h**) as-deposited, (**i**) 400 °C annealed

bright-field TEM images. The lattice structure was clearly observed along the interface in the high-resolution TEM image as shown in Fig. 7.3c, indicating that only moderate diffusion occurred during the low-temperature annealing for multilayers with thin layer. For $\lambda = 200$ nm case, no significant microstructure modification was observed even after annealing up to 300 °C as shown in Fig. 7.3e. After high temperature annealing, a fully intermixed and completely alloyed structure was previously observed to replace the layered structure for thin layer case [29–31], and the aforementioned XRD results in Fig. 7.1c confirmed the complete alloying process after high temperature annealing in current work. In the $\lambda = 25$ nm case, partial disintegration of layered structure was observed after 400 °C annealing, while the layered structure was still maintained after 500 °C annealing in the $\lambda = 200$ nm case, although obvious diffusion along the layer interface and grain boundaries, as well as recrystallized Ni grains and voids.

7.3.2 *Hardness Characterization*

7.3.2.1 Strengthening Mechanisms in As-Deposited Multilayers

Figure 7.4 shows the hardness as a function of individual layer thickness of as-deposited Ti/Ni multilayers with layer thickness ranging from 200 nm to 6 nm. The hardness increases with decreasing individual layer thickness down to 12 nm, and then decreases with further reduced layer thickness. Two different dislocation-mediated models were applied to fit the experimental results: a Hall–Petch model based on dislocation pile-up along interface for the range with layer thickness larger than 50 nm and a confined layer slip (CLS) model based on single dislocation bowing between interfaces used in the range with layer thickness further down to 12 nm. The Hall–Petch fitting is shown as in Fig. 7.4 (red line) by the following equation:

$$H = H_0 + K\lambda^{-1/2} \tag{7.1}$$

where H_0 is a constant, here $H_0 = 1.67\,\text{GPa}$, K the strengthening factor from layer thickness and $K = 20.8\ \text{GPa}\cdot\sqrt{\text{nm}}$, which can be used to predict the maximum hardness for the multilayer system via [32]:

$$K_\tau = \sqrt{\frac{\tau^* \mu b}{\pi(1-\nu)}} \tag{7.2}$$

where τ^* is the interface barrier strength for slip transmission, μ the shear modulus, b the Burgers vector, and ν the Poisson ratio. This yields a maximum hardness of 5.9 GPa, close to the experimental maximum hardness 6 GPa. Furthermore, the CLS fitting is also shown in Fig. 7.4 (blue line) and the modified CLS stress $\sigma_{CLS} = H/2.7$ is given as [32]:

$$\sigma_{CLS} = \frac{M\mu b sin\theta}{8\pi\lambda}\left(\frac{4-\nu}{1-\nu}\right)\ln\left(\frac{\alpha\lambda}{bsin\theta}\right) - \frac{f}{\lambda} + \frac{\mu b}{(1-\nu)L} \tag{7.3}$$

where M is the Taylor factor of 3.1, θ the angle between the slip plane and the interface, α the core cut-off parameter, f the interface stress with a typical value 2–3 J/m^2, and L the spacing of the interface dislocation array. Substituting $\mu = 76\,\text{GPa}$, $b = 0.25\,\text{nm}$, $\nu = 0.31$ for Ni, and taking $\alpha = 0.12, f = 3, \theta = 60°$ and $L = 25$, the CLS model can fit the experimental data well with λ from 75 to 12 nm. The good fitting of dislocation associated models as shown in Fig. 7.4 not only confirms

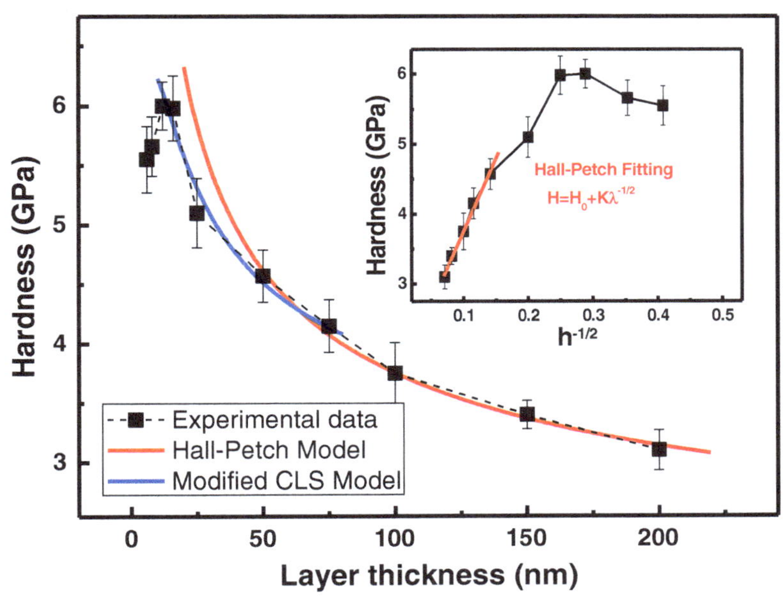

Fig. 7.4 Hardness results and dislocation-based modeling for as-deposited Ti/Ni multilayers: a Hall–Petch model for λ from 200 to 50 nm and a modified CLS model for λ from 75 to 12 nm

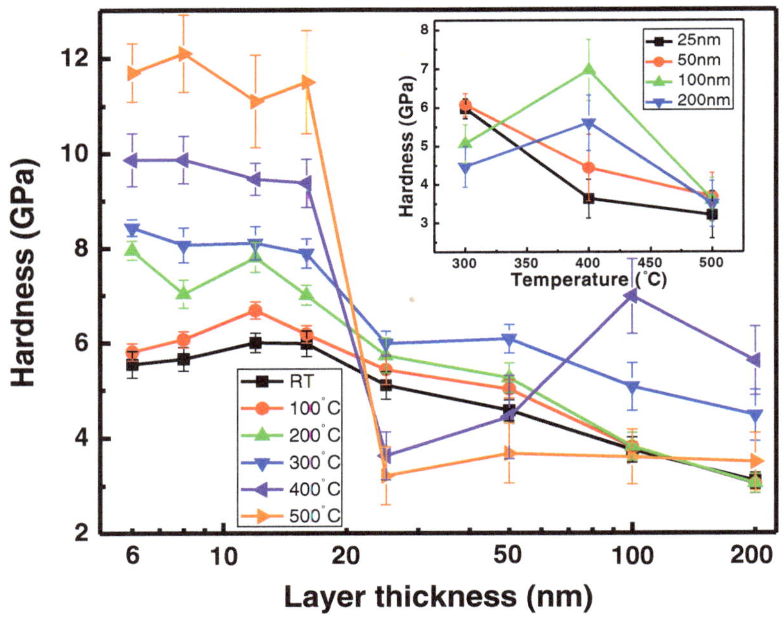

Fig. 7.5 Hardness results for both as-deposited and annealed Ti/Ni multilayer thin films. Insert is the hardness change after high temperature annealing for thick-layer case

the dislocation motion-dominated strengthening mechanisms, but also validates our experiments. The detailed analysis is also presented in our previous works [4, 28]. While the peak strength of 6 GPa was reached at $\lambda = 12$ nm, a decreased hardness was obtained with further decreasing λ. Both cross-section and surface morphology display polycrystalline structure in the thin-layer cases, and thus grain boundary mediated motions are considered as the probable dominating softening mechanism in this λ range.

7.3.2.2 Strengthening Mechanisms in Annealed Multilayers

Figure 7.5 shows the hardness for annealed Ti/Ni multilayers. Low temperature annealing is observed to enhance the multilayer strength due to grain boundary relaxation strengthening, which is a common annealing induced-strengthening mechanism for polycrystalline materials [32]. In the thin-layer case, the grain size is smaller and thus grain boundary occupies a larger volume portion. With more grain boundaries relaxed, there is more strengthening. With higher annealing temperature, there is also more strengthening, probably because higher temperature can provide sufficient energy to activate more grain boundary relaxation processes, leading to more strengthening. After 300 °C annealing, there are more obvious strengthening in the thick-layer cases, since grain boundary relaxation process has been activated during this period, leading to an initial strengthening. In the thin-layer cases, since a large extent of diffusion and even Ti-Ni intermetallic were detected, the solid solute strengthening and precipitation strengthening of intermetallic should be the dominating mechanisms instead. After high temperature annealing, the strong layer thickness dependent hardness evolution was shown. In the thick-layer cases, continued strengthening after 400 °C was observed due to grain boundary relaxation while further dramatic softening after 500 °C was due to grain growth. In the intermediate-layer cases, continuously decreased hardness was shown with increasing temperature also due to the grain growth, while the opposite trend in the thin-layer cases was resulted from the formation of Ti-Ni alloy.

7.3.2.3 Temperature-Thickness Dependent Strengthening Mechanism Map

The systematic study on coupled temperature-thickness effect on the hardness of Ti/Ni multilayers provides new insights into the strengthening/softening mechanism of nanometric metallic multilayers. A qualitative strengthening mechanism map is proposed for metallic multilayer system as shown in Fig. 7.6. This map covers all the possible strengthening/softening mechanisms with a wide range of temperature and layer thickness.

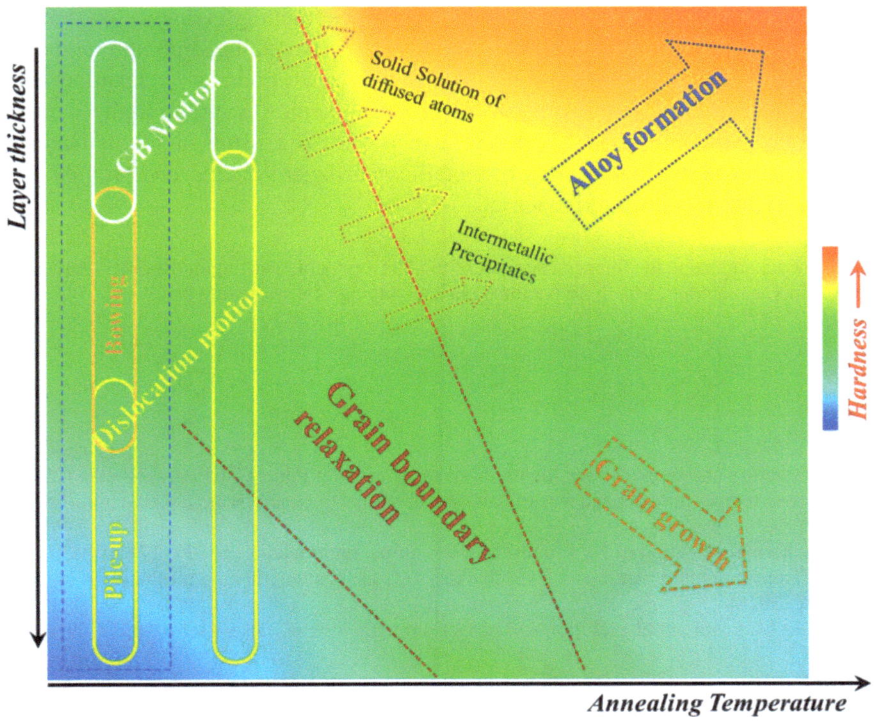

Fig. 7.6 Schematic of proposed strengthening/softening mechanism map of coupled temperature-thickness effect for Ti/Ni multilayers

For as-deposited multilayer thin films, dislocation-mediated motions are the dominating strengthening mechanisms with decreasing λ i.e. dislocation pile-up along interfaces with λ down to tens of nanometers and single dislocation transferring across interfaces with λ ranging from a few tens of nanometers to a few nanometers. With λ on the order of a few nanometers, the softening with further reduced λ is due to grain boundary-mediated motions. However, the grain boundary mediated motions can be suppressed by annealing-activated grain boundary relaxation process. In thin-layer cases, low temperature can trigger this grain boundary relaxation more easily, while higher temperature is required to provide sufficient energy in order to activate grain boundary relaxation in the thick-layer cases. In addition to grain boundary relaxation, this higher temperature also activate solid solution of diffused Ti and Ni atoms and precipitation of Ti-Ni intermetallics in the thin-layer cases. So the further strengthening in the thin-layer cases after higher temperature annealing is probably due to a combination of solid solution strengthening and precipitation strengthening of intermetallic. In the thick-layer cases, grain boundary relaxation-induced continuous strengthening is achieved even after high temperature annealing (e.g. 400 °C), while a dramatic hardness drop follows at further increased temperature (e.g. 500 °C) due to grain growth. In the intermediate-layer cases, high temperature-induced grain growth softening appears at a lower temperature (e.g. 400 °C). However, in the thin-layer cases, instead of grain growth, a fully intermixed and completely alloyed structure was formed, resulting in a continuously increased hardness with increasing annealing temperature. Note that the map in Fig. 7.6 only covers the effects from layer thickness and annealing temperature. An expanded map which also covers the loading orientation effect was reported in our recent work [4].

7.4 Conclusions

In the current work, a systematical study on coupled temperature-thickness effect on the strengthening mechanism of Ti/Ni multilayers was carried out with varying layer thickness from 200 to 6 nm and annealing temperature up to 500 °C. For thin-layer cases, a continuously increased hardness is observed with increasing temperature, while for intermediate-layer cases, a decreased hardness trend starts after high temperature annealing. For thick-layer cases, an increased hardness appears upon 300 °C annealing, continues after 400 °C annealing, and then a dramatic drop after further increased temperature (500 °C). These temperature-thickness dependent hardness behaviors in multilayer systems can be understood through microstructural characterization and analysis. For thin-layer cases, the continuous strengthening is due to low temperature-activated grain boundary relaxation, solid solution of diffused atoms, precipitation of intermetallic, and high temperature-induced alloying. For intermediate-layer cases, the initial strengthening is also due to grain boundary relaxation and the softening is due to recrystallization and grain growth. For thick-layer cases, the strengthening is due to higher temperature activated grain boundary relaxation, and the final softening is due to grain growth as well. The results highlight the temperature and thickness effect on the mechanical strength and thermal stability of metallic multilayer systems, and provide guidelines to control and optimize this important type of future engineered materials.

Acknowledgements The authors acknowledge the financial support of the University of Washington Regents. Part of this work was conducted at the UW Molecular Analysis Facility, a member of the NSF National Nanotechnology Infrastructure Network.

References

1. Ramos, A.S., et al.: Thermal stability of nanoscale metallic multilayers. Thin Solid Films **571**(Pt 2), 268–274 (2014)
2. Schoeppner, R.L., et al.: Elevated temperature dependence of hardness in tri-metallic nano-scale metallic multilayer systems. Thin Solid Films **571**(Pt 2), 247–252 (2014)
3. Cavaleiro, A.J., et al.: In-situ thermal evolution of Ni/Ti multilayer thin films. Intermetallics **51**, 11–17 (2014)
4. Yang, Z., Wang, J.: Coupled annealing temperature and layer thickness effect on strengthening mechanisms of Ti/Ni multilayer thin films. J. Mech. Phys. Solids **88**, 72–82 (2016)
5. Yang, Z., Stossel, M., Wang, J.: Microstructural evolution and surface strengthening of pulse-laser treated Ti/Ni multilayer thin films. Extreme Mech. Lett. **4**, 45–51 (2015)
6. Yang, Z., Lian, J., Wang, J.: Molecular dynamics simulation of thin film interfacial strength dependency on lattice mismatch. Thin Solid Films **537**, 190–197 (2013)
7. El-Awady, J.A.: Unravelling the physics of size-dependent dislocation-mediated plasticity. Nat. Commun. **6**, 5926 (2015)
8. Fan, H., et al.: Orientation influence on grain size effects in ultrafine-grained magnesium. Scr. Mater. **97**, 25–28 (2015)
9. Liu, Y., et al.: Mechanical properties of highly textured Cu/Ni multilayers. Acta Mater. **59**(5), 1924–1933 (2011)
10. Misra, A., Kung, H.: Deformation behavior of nanostructured metallic multilayers. Adv. Eng. Mater. **3**(4), 217–222 (2001)
11. Wang, J., Misra, A.: An overview of interface-dominated deformation mechanisms in metallic multilayers. Curr. Opinion Solid State Mater. Sci. **15**(1), 20–28 (2011)
12. Rao, S.I., Hazzledine, P.M.: Atomistic simulations of dislocation-interface interactions in the Cu-Ni multilayer system. Philos. Mag. A **80**(9), 2011–2040 (2000)
13. Hoagland, R.G., et al.: On the strengthening effects of interfaces in multilayer fcc metallic composites. Philos. Mag. A **82**(4), 643–664 (2002)
14. Embury, J.D., Hirth, J.P.: On dislocation storage and the mechanical response of fine scale microstructures. Acta Metall. Mater. **42**(6), 2051–2056 (1994)
15. Anderson, P.M., Li, C.: Hall–Petch relations for multilayered materials. Nanostruct. Mater. **5**(3), 349–362 (1995)
16. Huang, H., Spaepen, F.: Tensile testing of free-standing Cu, Ag and Al thin films and Ag/Cu multilayers. Acta Mater. **48**(12), 3261–3269 (2000)
17. Kang, B.C., et al.: Bilayer thickness effects on nanoindentation behavior of Ag/Ni multilayers. Scr. Mater. **57**(8), 703–706 (2007)
18. Masumura, R.A., Hazzledine, P.M., Pande, C.S.: Yield stress of fine grained materials. Acta Mater. **46**(13), 4527–4534 (1998)
19. Zhai, Q., et al.: Characterization of high strength Cu/Ag multilayered composites. In: Materials Research Society Symposium—Proceedings, 1997
20. Misra, A., Hoagland, R.G.: Effects of elevated temperature annealing on the structure and hardness of copper/niobium nanolayered films. J. Mater. Res. **20**(8), 2046–2054 (2005)
21. Wen, S.P., et al.: Thermal stability of microstructure and mechanical properties of Ni/Ru multilayers. Surf. Coat. Technol. **202**(10), 2040–2046 (2008)
22. Lee, H.-J., et al.: Thermal stability of a Cu/Ta multilayer: an intriguing interfacial reaction. Acta Mater. **47**(15–16), 3965–3975 (1999)
23. Bobeth, M., et al.: Thermal stability of nanoscale Co/Cu multilayers. Mater. Res. Adv. Tech. **92**(7), 810–819 (2001)
24. Troche, P., et al.: Thermally driven shape instabilities of Nb/Cu multilayer structures: instability of Nb/Cu multilayers. Thin Solid Films **353**(1), 33–39 (1999)
25. Misra, A., Hoagland, R.G., Kung, H.: Thermal stability of self-supported nanolayered Cu/Nb films. Philos. Mag. **84**(10), 1021–1028 (2004)
26. Lu, Y.Y., et al.: Morphological and mechanical stability of HCP-based multilayer nanofilms at elevated temperatures. Surf. Coat. Technol. **275**, 142–147 (2015)
27. Oliver, W.C., Pharr, G.M.: An improved technique for determining hardness and elastic modulus using load and displacement sensing indentation experiments. J. Mater. Res. **7**(6), 1564–1583 (1992)
28. Yang, Z., Wang, J.: Orientation-dependent hardness in as-deposited and low-temperature annealed Ti/Ni multilayer thin films. J. Appl. Mech. **82**(1), 011008 (2015)
29. Bhatt, P., et al.: High temperature annealing effect on structural and magnetic properties of Ti/Ni multilayers. Appl. Surf. Sci. **253**(5), 2572–2580 (2006)
30. Petrović, S., et al.: Formation of intermetallic phase in Ni/Ti multilayer structure by ion implantation and thermal annealing. Intermetallics **25**, 27–33 (2012)

31. Cho, H., Kim, H.Y., Miyazaki, S.: Alloying process of sputter-deposited Ti/Ni multilayer thin films. Mater. Sci. Eng. A **438–440**, 699–702 (2006)
32. Misra, A., Hirth, J.P., Hoagland, R.G.: Length-scale-dependent deformation mechanisms in incoherent metallic multilayered composites. Acta Mater. **53**(18), 4817–4824 (2005)

Chapter 8
A Novel Microdevice for In Situ Study of Mechano-Electrochemical Behavior with Controlled Temperature

Wonmo Kang, Marriner Merrill, Collin Becker, and Corey Love

Abstract Nanostructured electrodes have shown great potential in the development of Li-ion batteries with higher energy and power densities and longer cycle life. A fundamental understanding of the mechano-electrochemical behavior during charging/discharging cycles is essential for optimal and reliable design. Previous work has utilized in situ experimental techniques in an electron microscope to directly visualize material response during the reaction cycles. Unfortunately, the present in situ test methods are limited to room temperature and, as a result, the effect of temperature on charging/discharging cycles is not well understood. These electrochemical processes are intrinsically temperature sensitive, particularly for nanostructured electrodes. Here we present a novel microdevice that allows high resolution in situ observation of mechano-electrochemical response of nanomaterials in a scanning electron microscope with controlled temperature. The microdevice consists of built-in microcircuits for concurrent heating and temperature measurement during in situ experiments. To demonstrate these unique capabilities, we present the design, microfabrication and thermal characterization of this new class of microdevice.

Keywords Si anode • MEMS • In situ testing • Lithiation • Li-ion battery • High temperature

8.1 Introduction

With rapidly increasing use of electric vehicles (EV), implantable medical devices, and portable devices, demand for lithium-ion batteries with higher energy and power densities and longer cycle life is greater than ever. One way to significantly improve battery performance over current technology is to use higher capacity anode materials in lithium-ion batteries. Among many material candidates, silicon is an attractive choice because it offers the highest theoretical charge capacity, possesses a low discharge potential, and is abundant in the Earth's crust. For example, the charge capacity of silicon anodes is ten times higher than commonly used graphite anodes. Therefore, use of silicon anodes would lead to the development of smaller, lighter lithium-ion batteries with much longer operation time. Such batteries will greatly impact applications ranging from electric vehicles to portable electronics. Despite their superb advantages, commercial implementation of silicon anodes is still limited because they undergo significant volume changes (300–400 %) during operation due to insertion and extraction of lithium ions, called lithiation/delithiation cycles. This significant volume expansion/contraction often results in mechanical failure of silicon anodes.

Recently, the volume expansion-induced failure of silicon anodes has been overcome by utilizing one dimensional silicon (1-D Si) nanostructures, e.g., nanowires, nanorods, and nanotubes, that can accommodate the large volume change of silicon anodes [1–4]. Mass production of these various 1-D Si anodes would benefit from a well-developed industrial infrastructure in micromanufacturing of silicon. For example, various techniques to fabricate silicon nanostructures have been developed including top-down (etching standard crystalline Si substrates) or bottom-up (vapor–solid–liquid synthesis) approaches [3].

W.K was the American Society for Engineering Education—Naval Research Laboratory (ASEE-NRL) postdoctoral fellow and is now supporting NRL as a contractor with Leidos Corporation in Arlington, VA

W. Kang (✉) • M. Merrill • C. Love
US Naval Research Laboratory, Washington, DC, USA
e-mail: wonmo.kang.ctr.ks@nrl.navy.mil

C. Becker
US Army Research Laboratory, Adelphi, MD, USA

L. Starman et al. (eds.), *Micro and Nanomechanics*, Conference Proceedings of the Society for Experimental Mechanics Series,
DOI 10.1007/978-3-319-42228-2_8

Owing to the small size of 1-D Si anodes, in situ experimental techniques in scanning and transmission electron microscopy (SEM and TEM) can be used to directly visualize the Li–Si alloying reaction during the lithiation/delithiation cycles at the nanoscale [5–9]. Such in situ studies have revealed important underlying mechanisms in this reaction, including phase evolution, anisotropic expansion and fracture behavior, atomic level details of the lithiation of crystalline Si, and the effect of stress on the kinetics of lithiation [2, 9–12].

Current in situ studies have been limited to room temperature mainly due to lack of in situ testing method at high temperatures. However, electrochemical lithiation is an intrinsically temperature sensitive process affecting capacity/power fade, pack electrical imbalance, and other performance aspects [13]. Also, the Li–Si alloying reaction takes different pathways at different temperatures [2]. Finally, recent in situ thermo-mechanical studies revealed that single crystal silicon has a size dependent brittle-to-ductile transition (BDT), showing considerable plastic deformation at lower temperatures with reduction of sample size [14, 15]. Due to this size dependent BDT, deformation mechanisms of 1-D Si anodes could be altered during the Li–Si alloying reaction with variation of temperature. Despite the importance of temperature, comprehensive and robust in situ experimental techniques with controlled temperature are not available. As a result, remarkably little is known about the effect of temperature on the Li–Si alloying reaction in 1-D Si anodes [16], and on the corresponding phase evolution and mechanical response.

In addition to the thermal effects on electrochemical-mechanical processes during lithiation, testing above room temperature would also open new capabilities in characterizing battery anodes at their service temperatures. Service temperatures are currently a significant obstacle to battery use in applications ranging from consumer electronics to military and the drilling industry, where operational temperatures above 100 °C are often required [17].

To address the critical limitations to current in situ testing techniques, we have developed a novel microdevice-based testing method that allows in situ investigation of thermo-mechano-electrochemical processes in battery materials. In the present manuscript, we describe the design, fabrication, and characterization of the microdevice. Then we discuss potential applications of the microdevice-based in situ testing method.

8.2 Microdevice for In Situ Thermo- Electrochemical Testing

The microdevice-based method allows in situ thermo-electrochemical testing of micro/nanoscale specimens in a scanning electron microscope (SEM). The three-dimensional (3-D) schematic of this novel method is shown in Fig. 8.1a, b. The microdevice consists of a Si frame, two SiO_2 layers, built-in electrical circuits (metallic), and electrical contact pads. The Si frame is coated with an SiO_2 layer to prevent electrical crosstalk between circuits, i.e., the sensor and the heater circuits in Fig. 8.1b. The electrical circuits are fabricated using microfabrication techniques and are designed for effective heating of micro/nano-scale specimens with concurrent temperature measurements during in situ characterization. The outer surface of the metallic circuits on the bottom-side is protected by a second SiO_2 layer to prevent possible oxidation of the films during high temperature experiments. During in situ characterization, anode specimens on a conductive substrate are mounted on the top-side of the device, Fig. 8.1c. The substrate is heated by the Joule heating using the built-in microheater while the temperature is monitored by measuring temperature-dependent resistance of the sensor circuit. For lithiation/delithiation

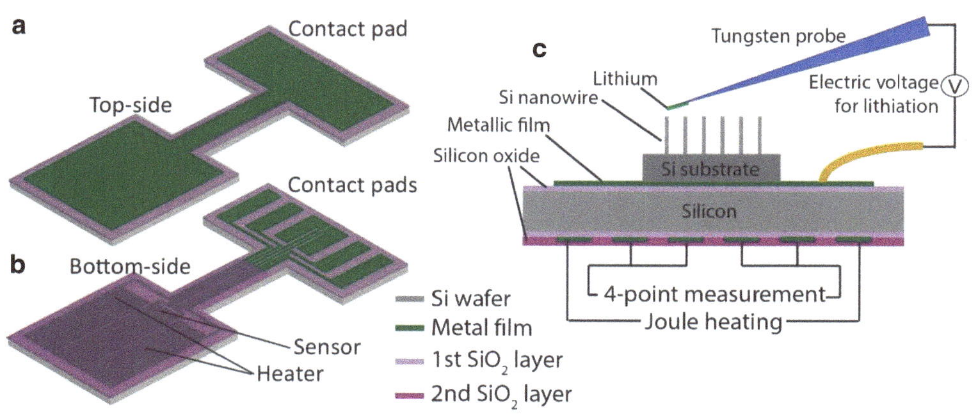

Fig. 8.1 Schematic of a microdevice for in situ testing of thermo-mechano-electrochemical response for micro/nano-scale specimens: (**a**, **b**) 3D schematics of the microdevice and (**c**) an overview of the in situ experimental setup for testing silicon anodes

cycles, each individual anode is brought into contact with a lithium-coated tungsten probe using a nanomanipulator and then an electrical potential is applied between the anode and the probe using an external power source.

Microfabrication Figure 8.2 shows the microfabrication procedure for the Si microdevice. First, the base SiO_2 layer (about 3 μm-thick) and a chromium thin film (about 200 nm-thick) were deposited on the both sides of a 500 μm-thick Si wafer using plasma-enhanced chemical vapor deposition (Oxford PECVD system) and sputter deposition (AJA system), respectively, shown in Fig. 8.2b. Then the first photoresist (PR, AZ 1518) layer was spin-coated on the Si wafer and patterned by photolithography (KARL SUSS MA6/BA6), Fig. 8.2c. For fabrication of the electrical circuits and contact pads, the patterned PR layers were used as a mask during a wet etch of the metallic layer, Fig. 8.2d. After removal of the first PR layers using acetone, the second PR layers (SPR220-7) were spin-coated and patterned to protect the device during inductively coupled plasma etching (Oxford ICP-RIE system), Fig. 8.2e. The silicon wafer was through-etched from the top to release individual free-standing devices, Fig. 8.2f. Again acetone was used to remove the second PR layers and then the second SiO_2 layer (about 1.5 μm-thick) was selectively deposited on the bottom-side to protect the sensor and heater, Fig. 8.2g. After completion of the microfabrication, the microdevices were annealed at 300 °C for 3 h in a vacuum chamber to minimize change in microstructures of chromium films during high temperature in situ experiments. Note that Si fames were fabricated at the wafer-level batch processing, which could be readily scaled up for mass production of the microdevices.

Experimental Setup For in situ SEM experiments, the microdevice is mounted on the device holder shown in Fig. 8.3a for ease of electrical connections and handling. This holder consists of a Macor (ceramic) frame, metallic contacts, a 6-pin pogo connector, and external electrical connectors. The current holder is designed to fit onto a sample stage of an FEI NOVA600 system, but can be easily modified for other SEM systems. The Macor frame is used to provide thermal and electrical isolation of the device from the holder. The metallic contacts and 6-pin pogo connector (not shown here) are for electrical connection to the top- and bottom-side of the device, respectively. Note that the spacing of the six contact pads matches with the 6-pin pogo connector. To ensure robust spring-loaded mechanical contacts between the pogo pins and pads,

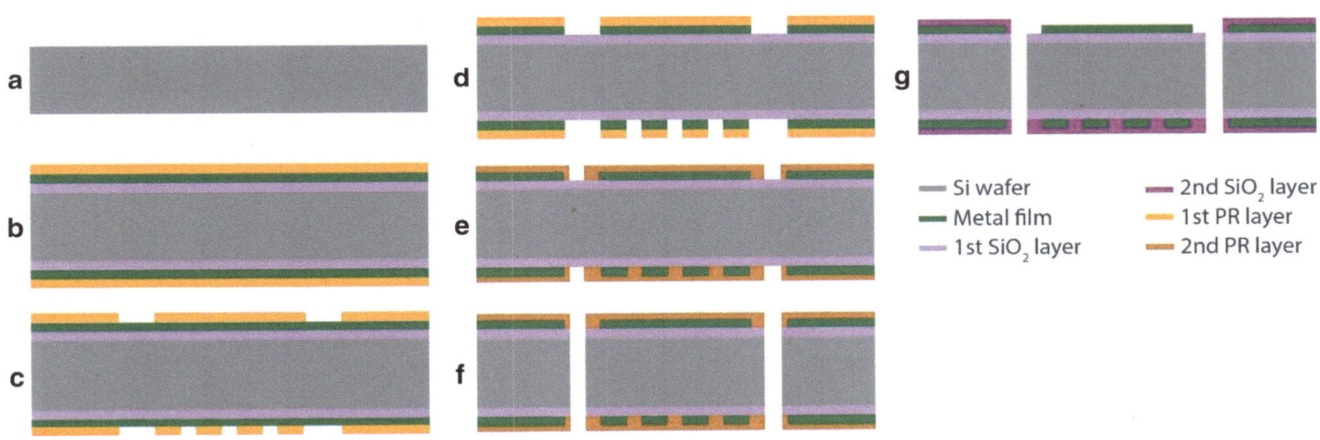

Fig. 8.2 A flow chart for fabrication of the microdevice using photolithography, deposition, and dry/wet etching techniques: (**a**) a Si wafer, (**b**) deposition of SiO_2 and chromium layers, (**c**) the first photolithography, (**d**) fabrication of electrical circuits and contact pads using wet etching, (**e**) the second photolithography, (**f**) anisotropic dry etching of the Si wafer, and (**g**) removal of the photoresist and deposition of the second SiO_2 layer

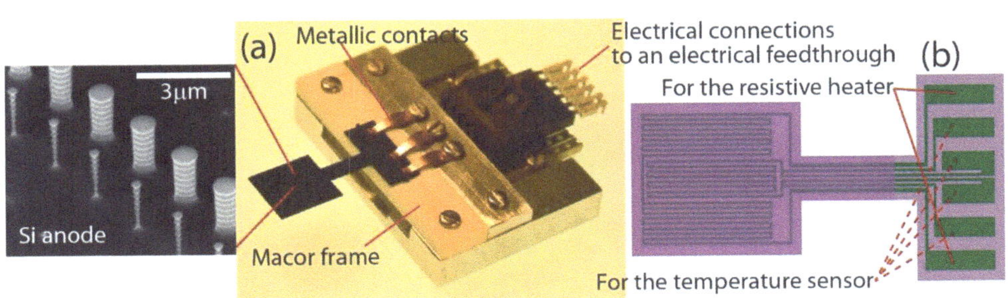

Fig. 8.3 Experimental setup for the microdevice-based in situ characterization method: (**a**) an optical image of the microdevice mounted on the device holder for in situ SEM experiments; the *inset* of (**a**) showing micro/nano-scale Si anodes; and (**b**) a detailed view of built-in electrical circuits on the bottom of the microdevice for heating of the anodes and concurrent temperature measurements

the metallic contacts apply gentle mechanical force on the top of the device as shown in Fig. 8.3a. The six contact pads are independently connected to the system electrical feedthrough through the six electrical pins to apply electrical inputs to the heater and monitor change in resistance of the sensor, respectively.

For heating of anode specimens, e.g., Si anodes mounted on the microdevice (the inset of Fig. 8.3a), an electrical potential is applied between the two electrodes connected to the resistive heater in Fig. 8.3b. For concurrent in situ temperature measurement, the temperature-dependent resistance of the sensor is monitored by four-point measurements using the four electrodes for the temperature sensor. In the following section, we experimentally characterize the thermal performance of the microdevice to demonstrate its unique capabilities to control temperatures during in situ material characterization.

8.3 Thermal Calibration

For temperature control, the microdevice utilizes the built-in sensor and heater that rely on temperature (T)-dependent resistivity (ρ) and resistive heating due to input voltage (V), respectively. Here we experimentally characterize their performance to achieve accurate temperature control during in situ experiments. In this regard, we first quantify the T-ρ correlation for the sensor and then use the sensor to characterize the heater, i.e., monitor T as a function of V.

To mimic in situ experimental conditions, the temperature sensor on the bottom of the device was calibrated in a vacuum chamber using a macroscopic heating device from MTII/Fullam with an aluminum spreader plate. The microdevice was directly placed on the plate with the sensor and heater up. A simple electrical connector with spring-loaded contacts was clamped down to the surface, providing both the electrical connection to the microdevice contacts as well as mechanically holding the microdevice in place. To characterize the temperature, four thermocouples were placed on the Al plate surrounding the microdevice and all within a few millimeters of its edges. An NI PXIe system was used to record the thermocouple data (PXIe-4353) and the four-point resistivity measurement (PXI-4071) of the built-in microsensor.

A series of thermal cycles were performed on the device with a combination of temperature ramps and holds, shown in Fig. 8.4. Only the average of the four thermocouples is shown because the difference in temperature between the hottest and

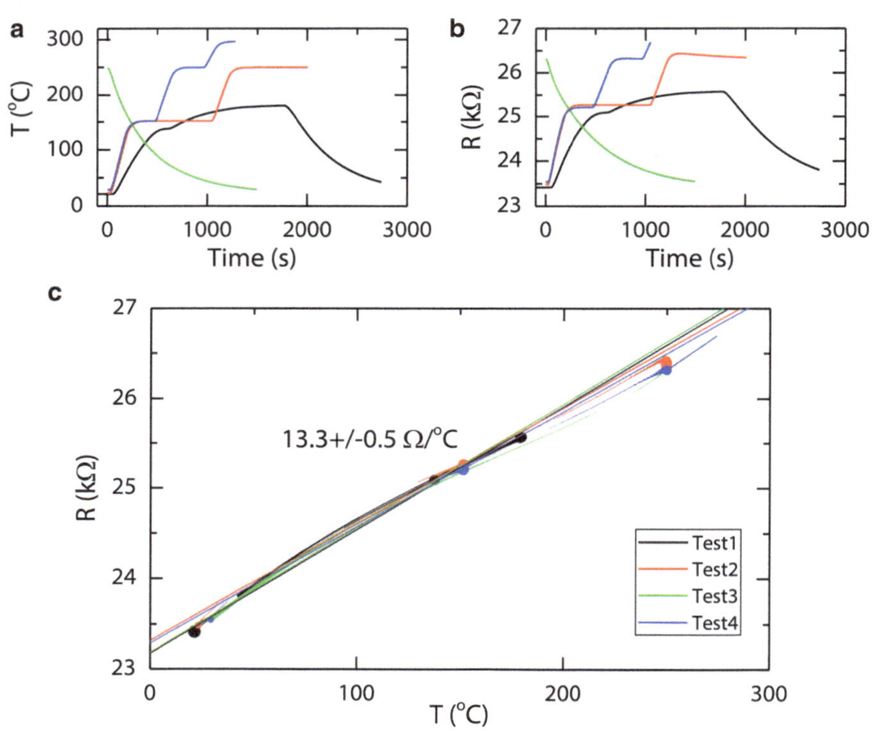

Fig. 8.4 Temperature (T)–Resistance (R) calibration over four tests. (**a**) Temperature over time. (**b**) Resistance over time. (**c**) Resistance vs temperature. In (**c**), data points are visually weighted using the derivative of the temperature to visually emphasize the temperature holds. No weighting was used in the calculation of the slope

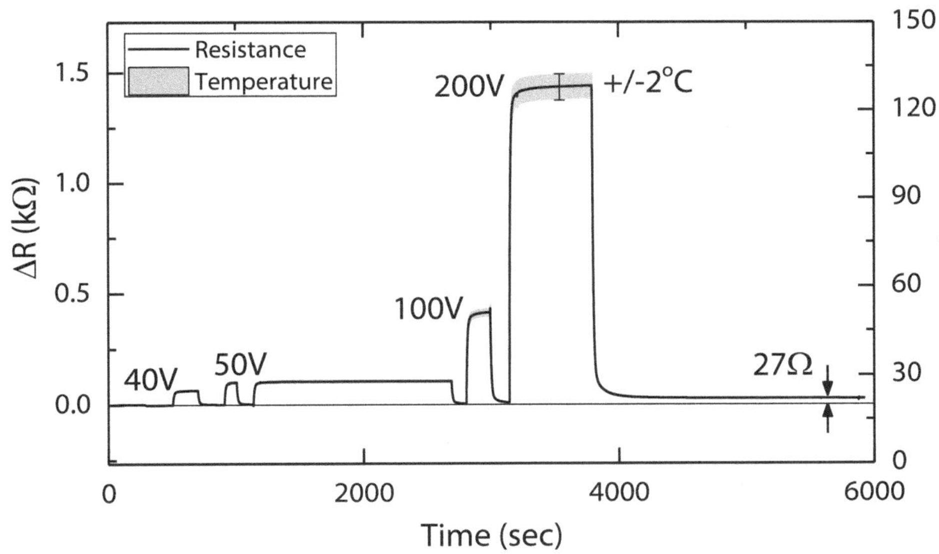

Fig. 8.5 Microheater verification. Voltage was applied as shown while measuring change in resistance (*left axis*). The calculated temperature is shown with its maximum error (*gray*) on the *right axis*

coolest thermocouples during tests had a maximum of 6 %. The change in resistance during temperature cycling is shown in Fig. 8.4b. Note that the resistance gradually decreased during the 250 °C holds in Test 2 and 4. This is likely due to either (1) deterioration in the electrical connector and solder connections or (2) microstructural changes in the metallic circuits due to further annealing. Although these can be easily addressed by either modifying the experimental setup or increasing initial annealing temperature during the microfabrication process, we concluded that the microdevice covers a sufficient range of temperatures for testing battery materials as they typically operate at <150 °C [17]. In Fig. 8.4c, linear fits of the four individual tests, 13.7, 12.9, 13.8, and 12.8 Ω/°C, are shown with an overall mean sensitivity of 13.3 ± 0.5 Ω/°C. Note that the linear fits were done using data only up to 160 °C as our main focus is on battery materials.

For verification of the heater, electrical input voltage was applied to the heater while the resistance of the sensor was measured to correlate the temperature with the voltage. The device was held in the designed holder as shown in Fig. 8.3 and an input voltage, 0–200 V (DC), was applied using a Keithley 2400. Current reached about 4 mA at the highest voltage. As before, the test was carried out in the vacuum chamber. The experimental results are shown in Fig. 8.5 where the temperature was calculated using the measured resistance and *R*–*T* relation, 13.3 Ω/°C. As shown, the coefficient uncertainty of ±0.5 Ω/°C resulted in a temperature uncertainty of ±2 °C at 120 °C (within 2 % error). The offset in the final resistance compared to initial (27 Ω) is negligible and demonstrates a lack of any significant change in the material properties.

8.4 Conclusion

We have developed a microdevice-based material testing method that allows in situ characterization of small scale materials during electrochemical processes, e.g., Li–Si systems, with controlled temperatures. Such unique in situ capabilities are essential to reveal the size- and temperature-dependent Li–Si alloy reaction in 1-D Si anodes and to map out a design parameter space for reliable and effective operation of lithium ion batteries in a wide range of temperatures. The microdevice was successfully fabricated, thermally calibrated, and then tested. Thermal calibration and verification testing showed good repeatability and use for temperatures under 150 °C, with a predicted thermal uncertainty of only a few degrees. We look forward to both the deeper fundamental understanding of the Si–Li system and also to further improving the capabilities of microdevice-based in situ testing techniques to enable even higher temperatures and expand to potential applications beyond battery materials.

Acknowledgements Funding for this project was provided by the Office of Naval Research (ONR) through the Naval Research Laboratory's Basic Research Program. During this project, W.K. was supported by the ASEE-NRL fellowship.

References

1. Arico, A.S., Bruce, P., Scrosati, B., Tarascon, J.M., Van Schalkwijk, W.: Nanostructured materials for advanced energy conversion and storage devices. Nat. Mater. **4**, 366–377 (2005)
2. McDowell, M.T., Lee, S.W., Nix, W.D., Cui, Y.: 25th Anniversary article: understanding the lithiation of silicon and other alloying anodes for lithium-ion batteries. Adv. Mater. **25**, 4966–4984 (2013)
3. Zamfir, M.R., Nguyen, H.T., Moyen, E., Lee, Y.H., Pribat, D.: Silicon nanowires for Li-based battery anodes: a review. J. Mater. Chem. A **1**, 9566–9586 (2013)
4. Song, T., Hu, L.B., Paik, U.: One-dimensional silicon nanostructures for Li ion batteries. J. Phys. Chem. Lett. **5**, 720–731 (2014)
5. Boles, S.T., Thompson, C.V., Kraft, O., Monig, R.: In situ tensile and creep testing of lithiated silicon nanowires. Appl. Phys. Lett. **103**, 263906 (2013)
6. Gu, M., Li, Y., Li, X.L., Hu, S.Y., Zhang, X.W., Xu, W., Thevuthasan, S., Baer, D.R., Zhang, J.G., Liu, J., Wang, C.M.: In situ TEM study of lithiation behavior of silicon nanoparticles attached to and embedded in a carbon matrix. ACS Nano **6**, 8439–8447 (2012)
7. Liu, X.H., Wang, J.W., Huang, S., Fan, F.F., Huang, X., Liu, Y., Krylyuk, S., Yoo, J., Dayeh, S.A., Davydov, A.V., Mao, S.X., Picraux, S.T., Zhang, S.L., Li, J., Zhu, T., Huang, J.Y.: In situ atomic-scale imaging of electrochemical lithiation in silicon. Nat. Nanotechnol. **7**, 749–756 (2012)
8. McDowell, M.T., Lee, S.W., Harris, J.T., Korgel, B.A., Wang, C.M., Nix, W.D., Cui, Y.: In situ TEM of two-phase lithiation of amorphous silicon nanospheres. Nano Lett. **13**, 758–764 (2013)
9. McDowell, M.T., Ryu, I., Lee, S.W., Wang, C.M., Nix, W.D., Cui, Y.: Studying the kinetics of crystalline silicon nanoparticle lithiation with in situ transmission electron microscopy. Adv. Mater. **24**, 6034–6041 (2012)
10. Lee, S.W., McDowell, M.T., Berla, L.A., Nix, W.D., Cui, Y.: Fracture of crystalline silicon nanopillars during electrochemical lithium insertion. Proc. Natl. Acad. Sci. U. S. A. **109**, 4080–4085 (2012)
11. Ryu, I., Choi, J.W., Cui, Y., Nix, W.D.: Size-dependent fracture of Si nanowire battery anodes. J. Mech. Phys. Solids **59**, 1717–1730 (2011)
12. Ryu, I., Lee, S.W., Gao, H.J., Cui, Y., Nix, W.D.: Microscopic model for fracture of crystalline Si nanopillars during lithiation. J. Power Sources **255**, 274–282 (2014)
13. Bandhauer, T.M., Garimella, S., Fuller, T.F.: A critical review of thermal issues in lithium-ion batteries. J. Electrochem. Soc. **158**, R1–R25 (2011)
14. Kang, W.M., Saif, M.T.A.: In situ study of size and temperature dependent brittle-to-ductile transition in single crystal silicon. Adv. Funct. Mater. **23**, 713–719 (2013)
15. Chang, T.H., Zhu, Y.: A microelectromechanical system for thermomechanical testing of nanostructures. Appl. Phys. Lett. **103**, 263114 (2013)
16. Chan, C.K., Peng, H.L., Liu, G., McIlwrath, K., Zhang, X.F., Huggins, R.A., Cui, Y.: High-performance lithium battery anodes using silicon nanowires. Nat. Nanotechnol. **3**, 31–35 (2008)
17. Tikhonov, K., Koch, V.R.: Li-Ion Battery Electrolytes Designed for a Wide Temperature Range. Covalent Associates, Inc., Woburn (2006)

Chapter 9
High-Rate Micro-Compression Using an Elastic Half-Space Loading Configuration

Daniel Casem, Jeffrey Lloyd, and George Gazonas

Abstract Recent developments in high rate micro-mechanical testing have used miniaturized Kolsky (Split-Hopkinson) Bars as the basic loading technique. These methods, which employ optical instrumentation instead of strain gages, have been used to test samples in the 20–50 μm size range at rates as high as 1 M/s. However, difficulties attributed to machining and alignment of such small systems make higher strain-rates difficult to achieve. In this work, we investigate an alternate method in which a small compression sample will be loaded rapidly against a transparent elastic half space. As the specimen deforms, it exerts what is essentially a point load on the half space, resulting in an indentation at the contact point. This indentation is measured with a displacement interferometer and in principle can be used to calculate the force applied to the sample. This can then be used to determine the stress–strain response of the specimen. Because loading plates can be made and aligned to great precision, this method has the potential to reach strain-rates as high as 3 M/s provided the force can be correlated to measured displacement with sufficient accuracy. This paper presents preliminary experimental and numerical results that illustrate the challenges in implementing such an approach.

Keywords Kolsky bar • High strain-rate testing • Split Hopkinson pressure bar

9.1 Introduction

Recently, miniature or "micro" Kolsky bar techniques (μKB) have been developed to permit high rate compression of small (25–50 μm) metallic samples at very high (10 k/s–1 M/s) strain-rates [1, 2]. These techniques involve optical instrumentation of very small bars, in some cases as small as 127 μm diameter. However, difficulties associated with alignment and microstructural effects within the bars ultimately limit the strain-rates which can be obtained [3].

In the present work, we investigate an alternate configuration based on a point loaded elastic half space that may permit micro-compression at rates in the range of 1–3 M/s. Figure 9.1 shows such a half space. The displacement u(t) is related to the applied load P(t), and solutions to the transient problem are available [4]. Specific solutions can be found via finite element analysis (FEA).

This concept can be adapted to high rate compression testing as shown in Fig. 9.2. A small sample rests against a large transparent anvil plate and is impacted by an identical flyer plate. The anvil and flyer are large enough that for the duration of the experiment they appear semi-infinite. As a result of impact, the specimen deforms plastically, applying a localized force $P_2(t)$ to the anvil.

It is assumed that the specimen is small enough that it is in a state of equilibrium. Therefore the quasistatic assumption is made, i.e., $P_1(t) = P_2(t)$, and the same force (P_1 in the figure) acts between the flyer and specimen. The displacement u_2 is measured with an interferometer through the transparent anvil. The force P_2 can, in principle, be determined from u_2. By symmetry, the motion u_1 is found from the known impact speed and the measured u_2, i.e., $u_1(t) = v - u_2(t)$. Specimen stress and strain are then found by $\sigma = P_2/A_0$ and $\varepsilon = (u_1 - u_2)/L_0$ where A_0 and L_0 are the initial cross-sectional area and length of the sample. Again, in this is example the anvil plate and flyer remain elastic.

There are several advantages of this configuration over the μKB. First, flat, transparent anvil plates can be made and aligned very precisely using established plate impact techniques. A material such as single crystal sapphire can be used for plates to essentially eliminate any microstructural effects, even at very small length scales. However, the major disadvantage of this method is that the applied force must be determined from the measured displacement. At this point, this issue is unresolved. Although exact solutions exist to the problem of Fig. 9.1, it is not clear if they can be readily applied to determine

D. Casem (✉) • J. Lloyd • G. Gazonas
Weapons and Materials Research Directorate, US Army Research Laboratory, Aberdeen Proving Ground, Aberdeen, MD 21005, USA
e-mail: daniel.t.casem.civ@mail.mil

L. Starman et al. (eds.), *Micro and Nanomechanics*, Conference Proceedings of the Society for Experimental Mechanics Series,
DOI 10.1007/978-3-319-42228-2_9

Fig. 9.1 An elastic half space loaded by a point load

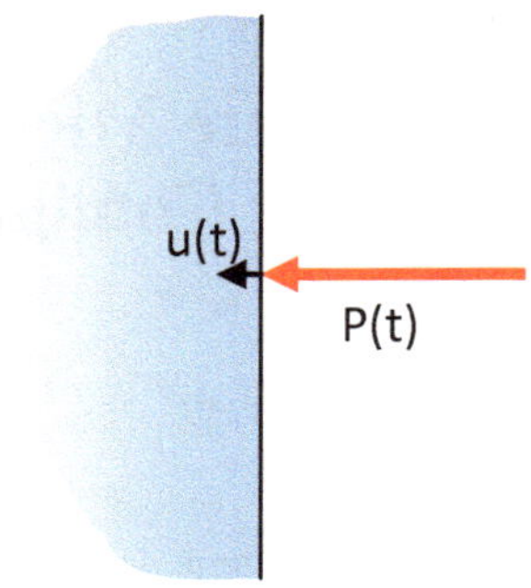

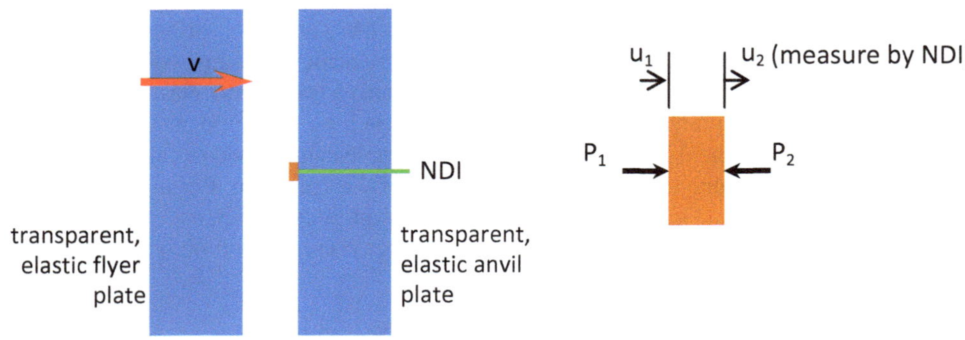

Fig. 9.2 Proposed configuration for a high rate experiment. A small compression sample is loaded between two elastic plates. The deformation of one plate is monitored with an interferometer and used to determine the response of the sample

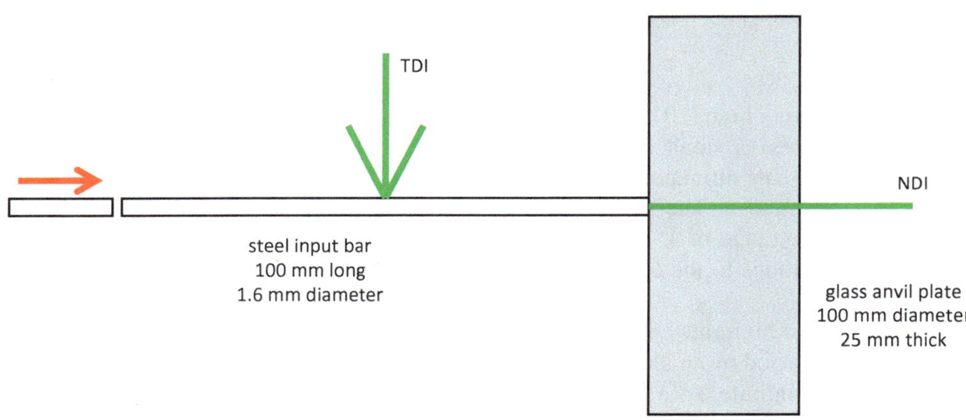

Fig. 9.3 An input bar is used to apply a transient "point-load" to an elastic anvil. The motion of the interface is monitored through the anvil using an NDI. It can also be calculated from the TDI measurements of the elastic waves in the input bar using a traditional Kolsky bar analysis

the applied transient load from a measured displacement. An inverse approach using a finite element analysis is also a possibility, and is currently under investigation. The remainder of this paper describes early experimental and numerical results that investigate the feasibility of instrumenting the elastic half space in such a way that the required displacement measurements can be made, as well as a summary of other issues and challenges that need to be addressed in developing this technique.

9.2 Results

In this section we demonstrate a measurement of contact point motion. The experiment involves replacing the output bar in a miniature Kolsky bar with a half space made from borosilicate glass, as shown in Fig. 9.3.

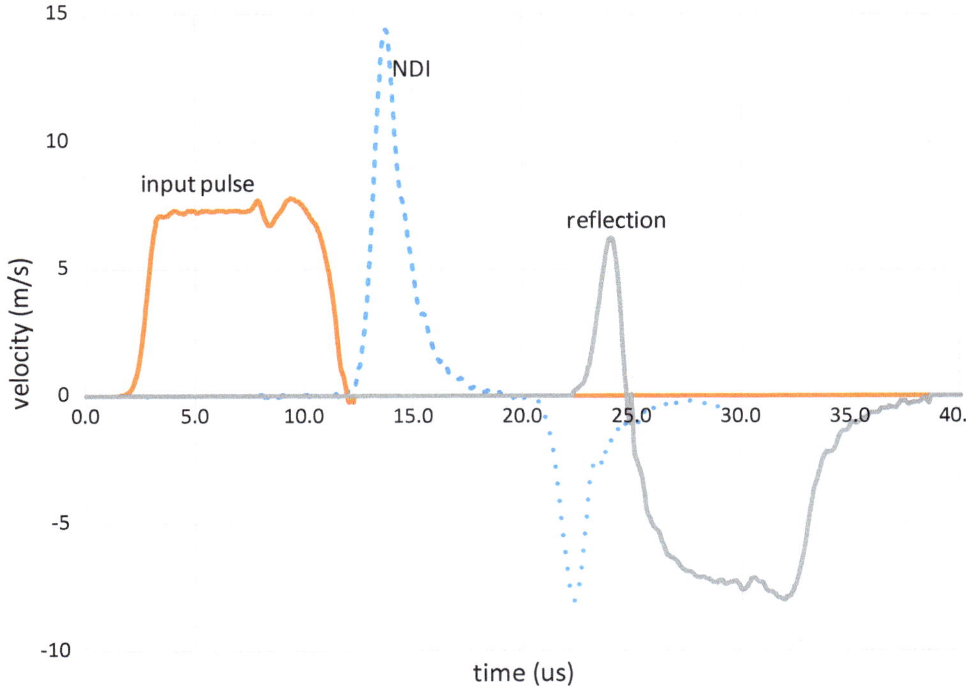

Fig. 9.4 Input pulse and reflection, in terms of particle velocity, measured by the TDI. Also shown is the velocity of the contact point measured directly by the NDI

A layer of aluminum is deposited on the contact side of the glass to provide a mirrored surface. No specimen is used, and the input bar is initially in contact with the plate. The input bar is steel, 100 mm long, and 1.6 mm in diameter. A transverse displacement interferometer [5], or TDI, is used to measure the particle velocity at the mid-point of the input bar as described in [1, 2]. The glass is 100 mm in diameter and 25 mm thick. A normal displacement interferometer (NDI) is used to measure the motion of the contact point. Assuming a wave speed of approximately 5000 m/s in the glass, there are approximately 10 μs after the glass is initially loaded until reflections from the back surface of the glass reach the measurement location, i.e., for 10 μs we can assume the glass is semi-infinite. A 22 mm long striker is used, which generates a pulse just under 10 μs long.

The input pulse magnitude is 7.5 m/s. A small amount of grease is used as a pulse shaper to reduce high frequency components in the input pulse. The contact point was inspected under a low-power microscope after the experiment to ensure that no obvious inelastic deformation occurred. The motion of the contact point is measured directly by the NDI, and can also be determined from an analysis of the input and reflected waves in the input bar using a standard Kolsky bar method. Figure 9.4 shows the measured particle velocity profiles (input pulse, reflection, and NDI output).

The measured signals are as expected. The input pulse is roughly square.[1] The reflection is initially positive, although as the glass is loaded it applies a larger force to the end of the input bar and the reflection becomes negative (the input bar rebounds from the glass anvil). The NDI shows a similar behavior: initially the motion is positive and holds steady once it is fully loaded. The motion then reverses as the glass unloads.

One main objective of this experiment is to confirm that the NDI can make a reliable measurement of contact point motion through the glass. There is some question as to whether the changing index of refraction of the stressed glass will affect this measurement. However this does not seem to be the case here. The motion of the contact point measured by the NDI is compared to that from the Kolsky bar analysis in Fig. 9.5, and they are in good agreement.

Although the measurement using the NDI at the applied load (i.e., directly under the sample in Fig. 9.1) seems a logical choice, it is not without other problems. For one, there is the question of whether or not the sample will be heated by the laser needed by the interferometer. In addition, at least for interferometers that require a specular surface as the NDI used here,

[1]Note the input pulse exhibits some oscillations at about 8 μs. These are due to flexural waves in the input bar due to a non-centric impact of the projectile. Multiple TDI measurements can be used to cancel these bending wave components from the velocity measurement [6, 7]; however for these initial experiments the additional instrumentation was omitted for simplicity.

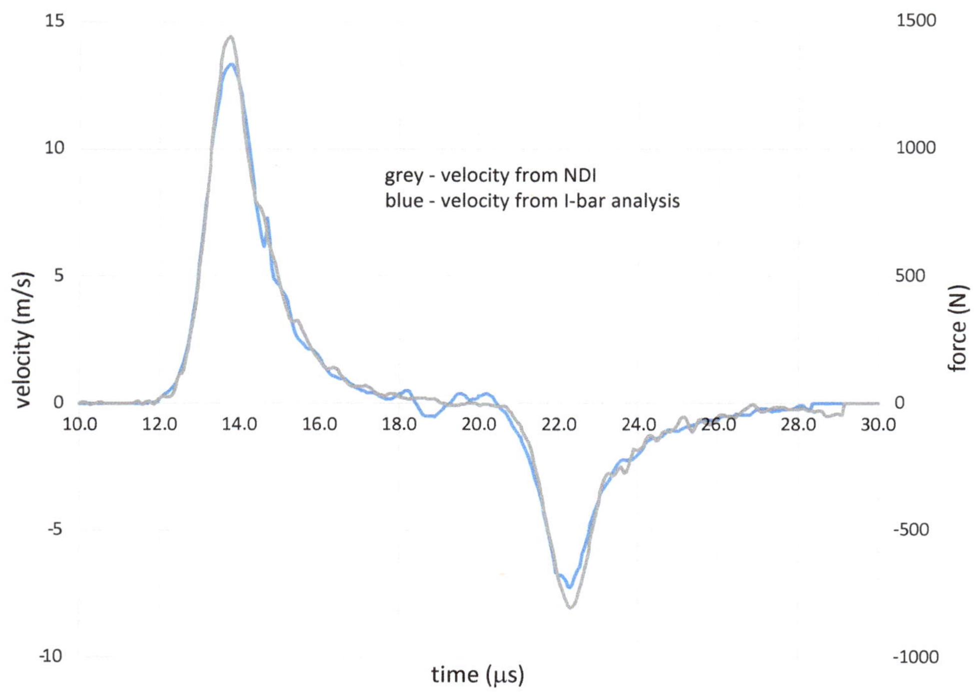

Fig. 9.5 Velocity of the contact point measured by the NDI and by the Kolsky bar analysis of the waves in the input bar

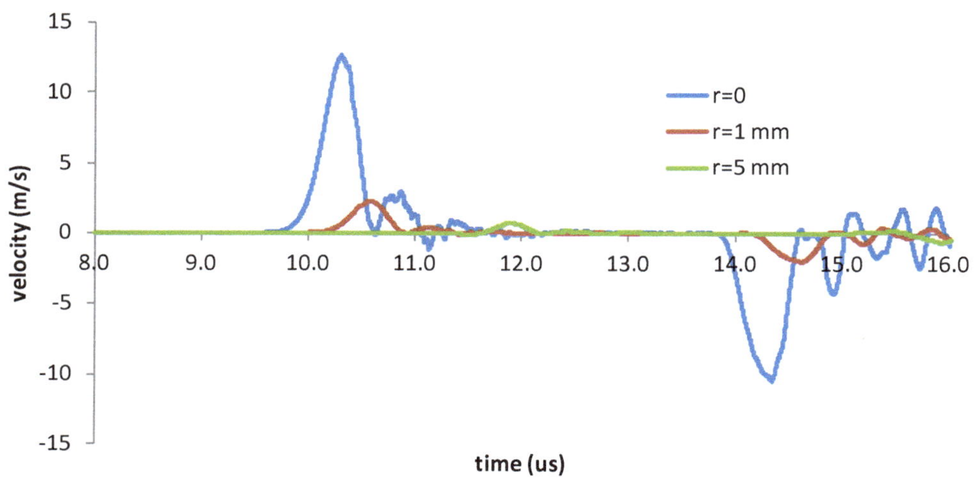

Fig. 9.6 Results from a simulation of the experiment shown in Fig. 9.3. Velocity of the glass interface is shown at three different radial locations. Note how rapidly the magnitude decays

there is the issue of whether or not the reflective properties of the surface will degrade when loaded by a rapidly deforming sample. Although in theory we could choose to make our measurements at a wide variety of locations on the anvil plate, it is well known for problems of this type the magnitude of the elastic wave decays rapidly away from the point of load application. This is demonstrated in Fig. 9.6, which shows results from a simulation of the experiment shown in Fig. 9.3. The velocities at several radial locations are shown, and the decay is clear. Thus instrumentation at another location may not be practical in many cases.

9.3 Discussion and Conclusion

In this work, we have discussed preliminary results of an investigation into whether an elastic half space subjected to a transient normal point load can be used as a reliable transducer for measurement of that applied load. We have suggested the use of the NDI to measure the motion of the half-space under the point-load. As mentioned previously, the primary issue that needs addressing at present is that of relating measured displacement to applied load. An exact solution is given in [4] for the application of a Heaviside step load on an isotropic elastic half-space (also assuming the Lamé constant and the shear modulus are equal). However, it is not evident that this solution can be easily applied to our case of arbitrary loads in a more general, likely anisotropic (e.g., single crystal sapphire), application, where we seek to determine force from a known displacement.

As an alternate approach, finite element methods can be applied to this problem. By incorporating such a solution in an optimization scheme, one can use an inverse approach to determine the applied load P(t) such that the numerically determined displacement matches that measured in the experiment. This approach of course suffers from the disadvantages of non-uniqueness and that a functional form for the applied load must be assumed in advance. This latter point is equivalent to assuming a constitutive model for the sample, and seeking parameters to that model in some iterative solution using FEA. This however may be adequate for specimens that are reasonably well-behaved. It also removes the restriction that the plate be semi-infinite, since the numerical solution can handle more complex boundary conditions. Thus one could possibly use a thinner plate and measure motion of the free-surface. This would eliminate several practical concerns, one of which would be the need for a transparent anvil plate. With this latter requirement, the method would be limited to fairly soft materials that would not damage the loading plates.

Finally, we note that although the intended application of this method is for high rate micro-mechanical compression and an extension of μKB methods, it seems likely that other applications exist in the field of high-rate solid mechanics. These include high-rate indentation, or possibly measurements of impact force by a Taylor impact test.

References

1. Casem, D.T.: A small diameter Kolsky bar for high-rate compression. In: Proceedings of the 2009 SEM Annual Conference and Exposition on Experimental and Applied Mechanics, Albuquerque, 1–4 June 2009
2. Casem, D.T., Grunschel, S.E., Schuster, B.: Normal and transverse displacement interferometers applied to small diameter Kolsky bars. Exp. Mech. **52**, 173–184 (2012)
3. Casem, D., Huskins, E., Ligda, J., Schuster, B.: A Kolsky bar for high-rate micro-compression – preliminary results. In: Proceedings of the 2015 SEM Annual Conference & Exposition (2015)
4. Pekeris, C.L.: The seismic surface pulse. Geophysics **41**, 469–480 (1955)
5. Kim, K., Suk, X., Clifton, R.J., Kumar, P.: A combined normal- and transverse-displacement interferometer with an application to impact of y-cut quartz. J. Appl. Phys. **48**(10), 4132–4139 (1977)
6. Casem, D.T., Zellner, M.: Kolsky bar wave separation using a photon Doppler velocimeter. Exp. Mech. **53**, 1467–1473 (2013)
7. Huskins, E.L., Casem, D.T.: Compensation of bending waves in an optically instrumented miniature Kolsky bar. J. Dyn. Behav. Mater. **1**, 65–69 (2015)

Chapter 10
Broadband Electromechanical Spectroscopy: A Method for Measuring the Dynamic Electromechanical Response of Ferroelectrics

Charles S. Wojnar, Jean-Briac le Graverend, and Dennis M. Kochmann

Abstract To gain a better understanding of the dynamic response of thermo-electro-mechanically coupled materials, in particular stiffness and mechanical damping, an experimental apparatus and method called Broadband Electromechanical Spectroscopy (BES) has been developed. The motivation for creating BES was to study the behavior of ferroelectric materials (and other electro-active materials) over wide ranges of frequencies of simultaneously-applied electric fields and mechanical stresses under accurate temperature control. By precisely controlling electric fields and mechanical loading, the effect of microstructural changes, in particular domain switching in ferroelectrics, on the material's dynamic mechanical response is measured. Experiments show large increases in mechanical damping during domain switching that can be tuned by appropriate electrical loading. Results obtained using the new capabilities brought by BES can be used to better understand the damping associated with domain wall motion to be able to create ceramic materials with both high stiffness and high damping—an elusive combination of properties in typical engineering materials. The design and use of the apparatus along with results obtained from tests on a particular ferroelectric, viz. lead zirconate titanate, are presented.

Keywords Dynamic mechanical analysis • Electromechanical loading • Ferroelectricity • Lead zirconate titanate • Viscoelasticity

10.1 Introduction

In engineering materials and structures it is important to minimize mechanical vibration. For example, reducing vibrations in aircraft structures is important for reducing fatigue and noise. In spacecraft structures, vibration mitigation is important for protecting payloads and improving the accuracy of scientific instruments. In addition, it is also useful for structural materials to have high stiffness to be able to support large loads. Unfortunately, naturally occurring materials exhibit either high damping and low stiffness (e.g. polymers) or low damping and high stiffness (e.g. ceramics). Nonetheless, a material exhibiting both high stiffness and high damping would be useful in a myriad of applications.

Various approaches can be taken to reduce vibrations. Passive methods involve combining high damping materials into structures, which results in a compromise of stiffness and damping. Alternatively, structural damping can be achieved by tuned-mass dampers. However the increased damping from such an approach is highly frequency dependent and its implementation adds weight, which is costly for aerospace applications. Active methods involve actuating structures in such a way to cancel out vibrations. This is typically achieved by piezoelectric actuators. Piezoelectric materials exhibit a linear relationship between applied stress and resulting charge accumulation on the surface (electric displacement). This phenomenon is used in shunted piezoelectrics where charge accumulation is dissipated by connecting surface electrodes with a resistor [1]. But this approach is also highly frequency dependent. This effect has been used in composites; piezoelectric inclusions are embedded in a metal matrix such that charge accumulation is dissipated via Joule heating [2], which only results in moderate damping increases. The converse relationship (linear relationship between applied electric field and

C.S. Wojnar (✉)
Mechanical and Aerospace Engineering, Missouri University of Science and Technology, 400 W. 13th St., Rolla, MO 65409, USA
e-mail: wojnarc@mst.edu

J.-B. le Graverend
Department of Aerospace Engineering, Texas A&M University, TAMU 3141, College Station, TX 77843, USA

D.M. Kochmann
Graduate Aerospace Laboratories, California Institute of Technology, 1200 E. California Blvd. MC 105-50, Pasadena, CA 91125, USA

L. Starman et al. (eds.), *Micro and Nanomechanics*, Conference Proceedings of the Society for Experimental Mechanics Series,
DOI 10.1007/978-3-319-42228-2_10

strain) is what is used in piezoelectric actuators where an applied electric field is carefully controlled to generate strains in such a way to cancel out vibrations. The drawback of this approach is that piezoelectric actuators exhibit small strains, which makes damping large deformations difficult.

The approaches using piezoelectrics, however, only utilize their linear response. Piezoelectrics like the widely used material lead zirconate titanate (PZT) are also ferroelectric. That is, they exhibit a spontaneous polarization that can be reoriented by applying sufficiently-large electric fields and stresses. The process of polarization reorientation, or so-called domain switching, is a dissipative process [3]. Nonetheless, there are only a limited number studies on using domain switching as an alternative mechanism for mechanical damping. In particular, [4] characterized the damping during electric field induced domain switching only for low mechanical frequencies. In addition, experiments were only performed in bending and have not been reported for other types of loading such as torsion. This is in part due to a lack of experimental methods. We close this gap by presenting an experimental setup and method called Broadband Electromechanical Spectroscopy (BES) for measuring the dynamic mechanical response (i.e. stiffness and damping) of ferroelectrics (and other materials) while applying sufficiently-large electric fields to induce domain switching. Using BES, the stiffness and damping of a particular ferroelectric, viz. PZT, was measured over a wide range of bending and torsional frequencies. A general overview of the method and a representative set of experimental results are presented. For a more detailed explanation of the setup and additional experimental results we refer the reader to [5, 6].

10.2 Materials and Methods

Although initial experiments are performed on PZT, which has a high Young modulus (typically 50 GPa), a method and apparatus was desired to test a wide range of specimen properties (including more compliant ferroelectric polymers such as polyvinylidene fluoride (PVDF) with a Young modulus of 3 GPa [7]) over many decades of frequency. Therefore, the apparatus developed for BES is based on Broadband Viscoelastic Spectroscopy (BVS) [8]. That is, cantilevered beam specimens are tested in bending and torsion using contact-less techniques. A magnet is attached to the free end of the specimen such that when placed between a pair of Helmholtz coils, the magnetic field from the coils applies a bending and torsional moment to the specimen as shown in Fig. 10.1. The resulting deflection and twist of the specimen is measured via a laser detector setup; an incoming laser beam reflects off of a mirror attached to the specimen's free end into a laser position

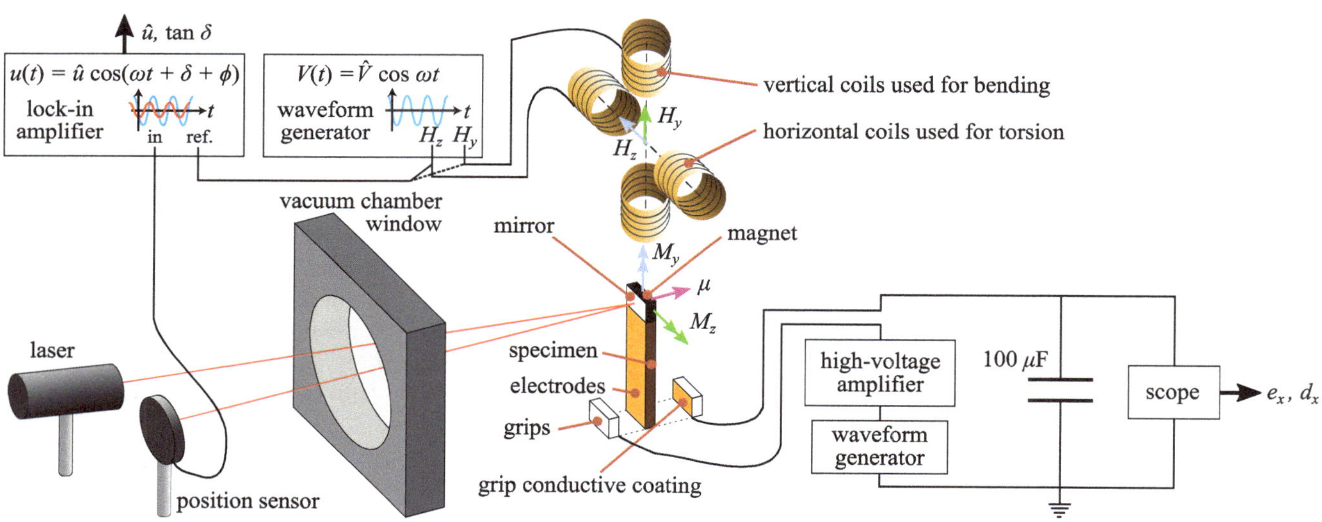

Fig. 10.1 Schematic of the BES apparatus (Adapted from [5]) showing a specimen gripped in the *center*. A vertical and horizontal pair of Helmholtz coils (shown in raised position) are used to apply bending and torsional moments, respectively, to the specimen through a permanent magnet attached to the specimen's free end. A mirror also attached to the specimen's free end reflects an incoming laser ray into a position sensor to measure the specimen's deflection and twist. The setup is placed inside a vacuum chamber with a window to allow for the laser beam to pass through. The *top-left* portion of the drawing shows the waveform generator used to send current through the coils and the lock-in amplifier used to measure the resulting amplitude and phase of the laser position. The *bottom-right* portion of the drawing shows the Sawyer-Tower circuit used for applying a voltage across the specimen's surface electrodes via conductive grips and for measuring changes in the macroscopic electric displacement

sensor. Thus, changes in the specimen's deflection angle and twisting angle are captured by the movement of the reflected laser beam in the sensor. Going beyond current BVS setups, we add the capability to simultaneously apply electric fields to specimens via surface electrodes while applying bending and torsional moments. Furthermore, the experimental setup is placed inside a vacuum chamber to reduce parasitic damping due to the surrounding air as well as to allow for accurate temperature control via radiant heaters. However, only the results performed in air at room temperature are presented here and the reader is referred to [5] for additional details on the vacuum chamber and heating system.

10.2.1 Force Control

Helmholtz coils were constructed by winding magnet wire around a stiff, non-conducting, non-magnetic core made of Macor. The apparatus consists of two pairs of Helmholtz coils. When current is passed through the vertical and horizontal pairs of coils, magnetic fields H_y and H_z are generated along their axis such that they apply a bending moment M_z or twisting moment M_y, respectively, to the magnet attached to the specimen as depicted in Fig. 10.1. In the current setup, the magnetic moment of the attached permanent magnet is oriented in such a way that the applied moments are related to the magnetic field by $M_y = \mu H_z$ and $M_z = \mu H_y$, where μ is the magnetic moment of the attached magnet. Using the equation for the magnetic field inside a solenoid due to an applied current, the applied moment can be related to the current in the coils by,

$$M_y = \mu \alpha_z n_z i_z, \qquad M_z = \mu \alpha_y n_y i_y, \tag{10.1}$$

where α is a correction factor for the deviation of the magnetic field from the ideal, infinitely long solenoid, n is the number of windings per unit length, and i is the current. Subscripts y and z refer to the parameters associated with the vertical and horizontal coils, respectively.

10.2.2 Measuring the Specimen's Deflection and Twist

The applied bending or torsional moments cause the reflected laser ray to move vertically or horizontally in the laser detector by an angle θ_z and θ_y, respectively. If the position sensor is placed a distance l away from the specimen, the distance the laser ray moves vertically or horizontally on the sensor is then $u_z = l \tan \theta_z$ and $u_y = l \tan \theta_y$. For small deflection/twisting angles, the relationship is approximately $u_z = l\theta_z$ and $u_y = l\theta_y$. By performing bending or torsion tests, the specimen's dynamic Young and shear moduli can be determined by considering the static solution for the deflection angle and twist at the free end due to an applied bending and torsional moment,

$$\theta_z = \frac{M_z L}{E I_z}, \qquad \theta_y = \frac{M_y L}{G J_y}, \tag{10.2}$$

respectively, where L is the free length of the specimen, E is Young's modulus, G is the shear modulus, I_z is the bending moment of inertia, and J_y is the torsional rigidity. By substituting the relationship between deflection and twisting angle and laser displacement, as well as relationship between applied current and moment in Eq. (10.1), the moduli in Eq. (10.2) can be expressed in terms of the applied current and laser position as,

$$E = \frac{\mu \alpha_z n_z i_z l}{u_z I_z} \equiv C_z \frac{i_z}{u_z}, \qquad G = \frac{\mu \alpha_y n_y i_y l}{u_y J_y} \equiv C_y \frac{i_y}{u_y}, \tag{10.3}$$

where C_z and C_y contain parameters associated with the apparatus and specimen geometry, which were held constant for the experiments. To obtain the dynamic stiffness and damping when applying a sinusoidal current through the coils, the correspondence principle is applied to Eq. (10.3), where the current i and laser position u are replaced by their complex-valued amplitudes. Application of the correspondence principle results in complex-valued Young and shear moduli:

E^* and G^*, respectively (see e.g. [9]). Thus the dynamic stiffness in bending and torsion are their respective absolute values, which correspond to the amplitudes of the current and laser position (denoted with hats),

$$|E^*| = C_z \frac{\hat{i}_z}{\hat{u}_z}, \qquad |G^*| = C_y \frac{\hat{i}_y}{\hat{u}_y}. \tag{10.4}$$

We chose to report the relative Young and shear moduli compared to their values at zero electric field, denoted with superscript 0 while fixing the applied current. Thus, the relative moduli are simply the ratio of the laser position amplitudes,

$$\frac{|E|}{|E^0|} = \frac{\hat{u}_z^0}{\hat{u}_z}, \qquad \frac{|G|}{|G^0|} = \frac{\hat{u}_y^0}{\hat{u}_y}. \tag{10.5}$$

The damping is measured by the loss tangent in bending ($\tan\delta_E$) and torsion ($\tan\delta_G$), which is the tangent of the argument of the complex Young and shear moduli, respectively,

$$\tan\delta_E = \tan\left[\arg\left(\frac{\hat{i}_z}{\hat{u}_z}\right)\right], \qquad \tan\delta_G = \tan\left[\arg\left(\frac{\hat{i}_y}{\hat{u}_y}\right)\right]. \tag{10.6}$$

The laser position amplitude and phase difference between applied current through the coils and resulting laser position are computed during experiments using a lock-in amplifier. It should be noted that Fig. 10.1 shows that the applied voltage to the Helmholtz coils is used as the reference for the lock-in amplifier. Therefore, the phase difference between the voltage applied to the coils and the resulting current (denoted by ϕ in Fig. 10.1) must be subtracted from the phase difference measured by the lock-in amplifier to obtain the phase shift due to the viscoelastic response of the specimen δ. This phase shift was measured a priori and used to correct the measurement obtained by the lock-in amplifier and the reader is referred to [5] for details. Furthermore, it should be emphasized that Eq. (10.2) is the static solution to the deflection and twisting angle of the specimen and does not include inertial effects. Thus, the relative moduli and loss tangents in Eqs. (10.5) and (10.6) only correspond to the material properties away from resonance where inertial effects become significant. Near resonance, Eqs. (10.5) and (10.6) correspond to the structural properties of the specimen.

10.2.3 Electric Field Control and Measuring Electric Displacement

Upon application of sufficiently-large electric fields, domain switching occurs in the material and gives rise to changes in the average (macroscopic) electric displacement. Thus, to interpret the mechanical response of the specimen, it is necessary to measure the electric displacement to detect when domain switching is occurring. Measuring the electric displacement of ferroelectrics under large, cyclic electric fields is accomplished using a Sawyer-Tower circuit [10] as shown in Fig. 10.1. A triangular waveform is amplified by a high-voltage amplifier and is applied to surface electrodes on the specimen. Changes in the specimen's electric displacement causes charge to accumulate on its surface, which is measured via a reference capacitor in series with the specimen. The specimens tested had a thickness $h = 1$ mm and surface area (width $b = 3$ mm and length $L = 38$ mm) $A = bL$. Therefore, the electric field through the thickness e_x is the applied voltage on the specimen V divided by the thickness: $e_x = V/h$. In addition, the electric displacement through the thickness d_x is the total charge Q on the specimen (measured by the reference capacitor) divided by the surface area of the electrodes: $d_x = Q/A$.

10.2.4 Materials

Initially-poled Navy-type II PZT specimens (PSI-5A4E from Piezo Systems Inc., Woburn, MA, USA) were tested. Specimen sizes of $1 \times 3 \times 38$ mm^3 were cut and coated with approximately 10 μm thick nickel electrodes by the supplier. Typical grain sizes in the specimens were on the order of 2 μm.

10.3 Experimental Results

Using BES, bending and torsion experiments were performed to measure the evolution of the dynamic Young and shear moduli, respectively, as well as their associated loss tangents while applying large, cyclic electric fields at room temperature. Mechanical bending and torsional frequencies were much higher than the applied cyclic electric fields (by at least an order of magnitude). Thus, measurements correspond to the incremental moduli and loss tangent (or the specimen's structural response near resonance) for different values of the applied electric field. The large, cyclic electric fields result in domain switching causing significant variations in the specimen's incremental moduli and loss tangent.

10.3.1 Bending Experiments

To measure the dynamic Young modulus and loss tangent, specimens were subjected to bending moments by using the vertical pair of coils in the BES setup. Various mechanical bending frequencies were tested from 25 to 1000 Hz using a 7.2 Vpp (peak-to-peak) amplitude sinusoidal voltage applied to the coils. Simultaneously, a slowly-varying triangle wave electric field with an amplitude of 1.95 MV/m was applied at 0.1 Hz for all experiments.

The variation of the electric displacement was recorded for the different mechanical bending frequencies and is shown in Fig. 10.2. The hysteresis of the electric displacement matches the well-known behavior observed in numerous previous static studies (e.g. [11]). For the different mechanical frequencies tested and the applied moment amplitude, there is no change in the evolution of the electric displacement when applying electric fields as the corresponding curves in Fig. 10.2 coincide. It is important to note that domain switching occurs near certain values of the electric field called the coercive field e_c, which is typically taken where the electric displacement changes sign. This is where domain switching causes large variations in the macroscopic electric displacement. For the specimens tested, the coercive fields in the positive and negative directions were 1.49 MV/m and −1.36 MV/m, respectively, as seen in Fig. 10.2.

The relative dynamic Young modulus was measured at the same mechanical bending frequencies while applying the same cyclic electric field and the results are shown in Fig. 10.3. The modulus varies linearly with electric field away from domain switching (i.e. away from the coercive field). This can be attributed to the electric field dependence of the elastic constants [12]. Further softening in the Young modulus during domain switching (e.g. by 60 % for case of 25 Hz in Fig. 10.3a) is due to anisotropy of the elastic constants [13]; domain switching causes reorientation of the crystal structure and changes the overall elastic moduli of the specimen. Finally, domain switching occurring during each mechanical cycle causes an additional apparent softening of the dynamic modulus.

The loss tangent in bending was measured for same set of mechanical loading frequencies and the results are shown in Fig. 10.4. Similar to the relative Young modulus, the loss tangent in bending varies linearly away from domain switching which can be attributed to the electric field dependence of the complex elastic constants [14]. There is a large increase in

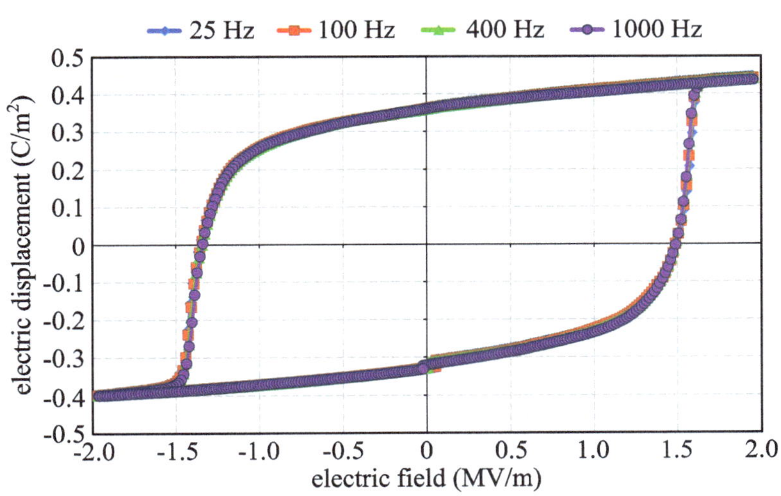

Fig. 10.2 Electric displacement of the specimen versus an applied triangle wave electric field at 0.1 Hz while simultaneously applying various mechanical bending frequencies of 25, 100, 400, and 1000 Hz (Adapted from [5])

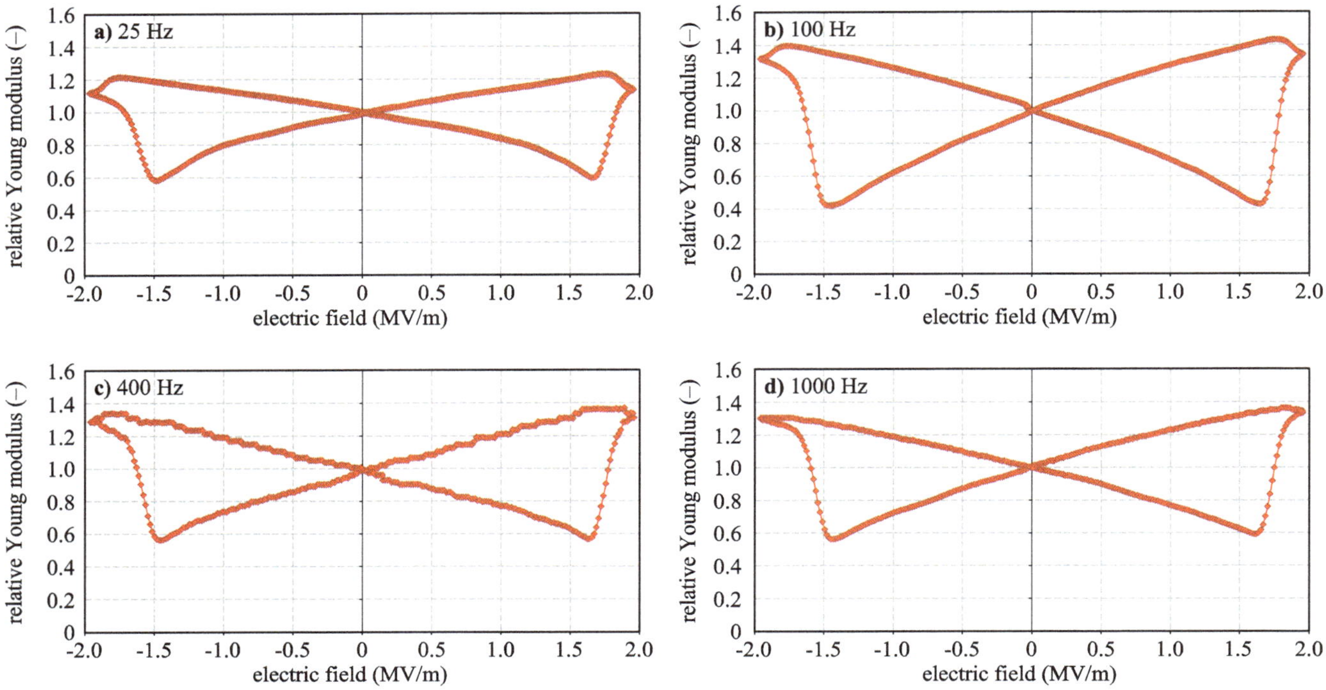

Fig. 10.3 Relative Young modulus versus an applied cyclic electric field at 0.1 Hz for mechanical bending frequencies of (**a**) 25 Hz, (**b**) 100 Hz, (**c**) 400 Hz, and (**d**) 1000 Hz. The Young modulus is normalized by the value at zero electric field (Adapted from [5])

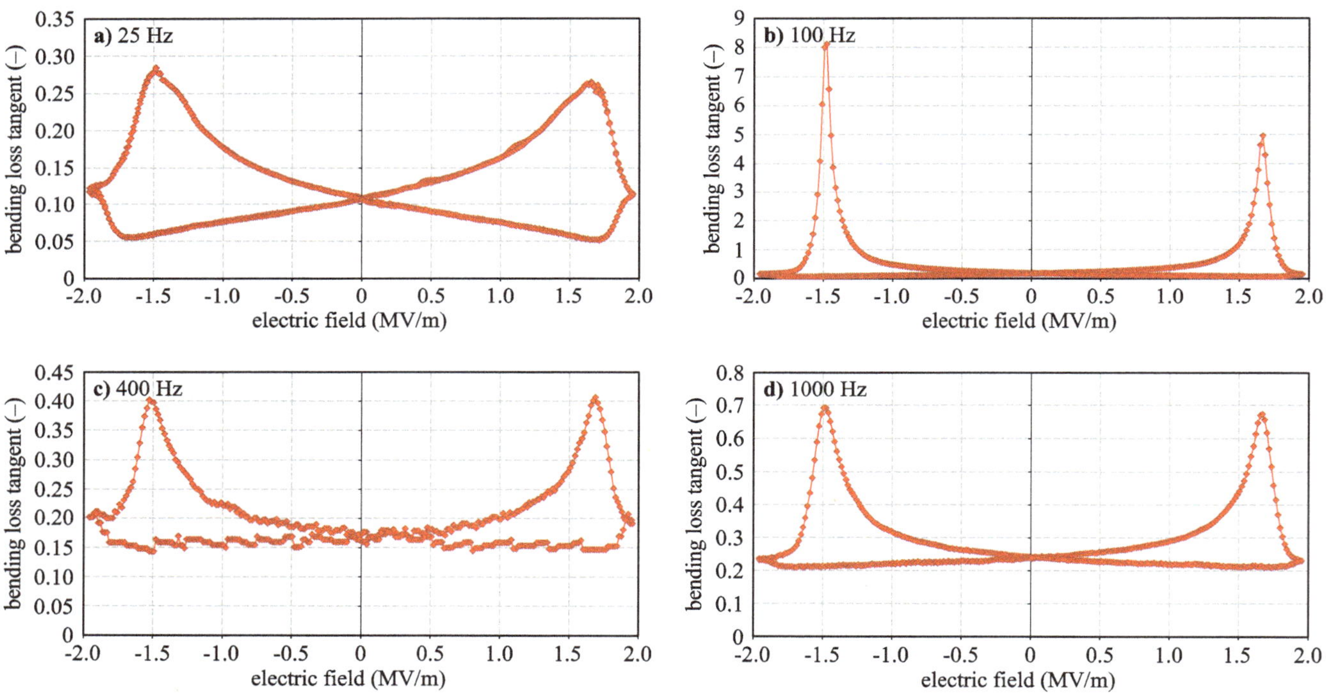

Fig. 10.4 Loss tangent in bending versus an applied cyclic electric field at 0.1 Hz for mechanical bending frequencies of (**a**) 25 Hz, (**b**) 100 Hz, (**c**) 400 Hz, and (**d**) 1000 Hz (Adapted from [5])

the loss tangent during domain switching (by nearly a factor of 3 for the case of 25 Hz in Fig. 10.4a). Typically, domain switching is accommodated by growing and shrinking individual domains by the motion of domain walls. Their motion dissipates energy [3] and causes an increase in the macroscopic loss tangent [6], which is the ratio of the energy dissipated

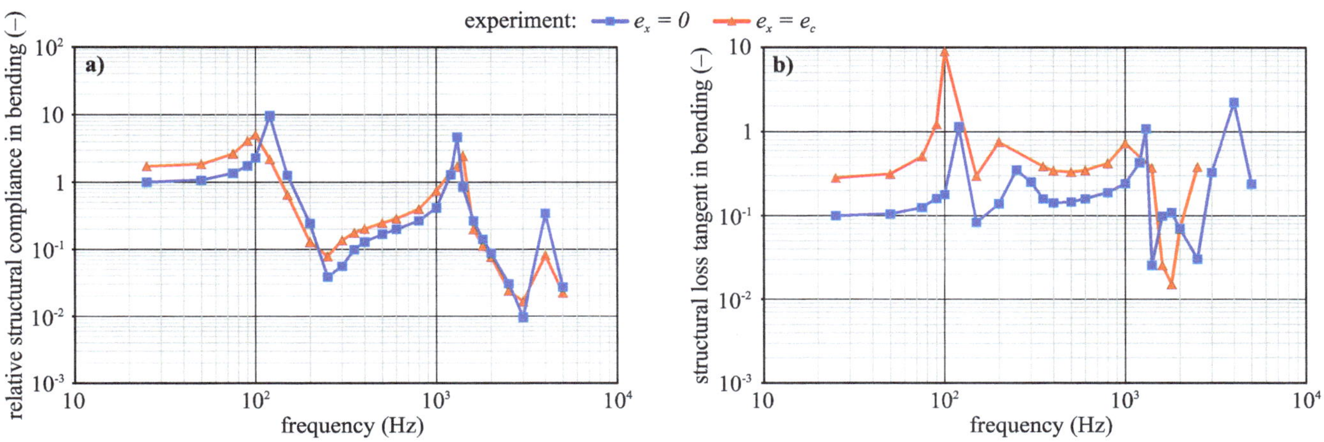

Fig. 10.5 (**a**) Relative compliance and (**b**) loss tangent versus bending frequency measured at zero electric field ($e_x = 0$) and at the coercive field ($e_x = e_c$) (Adapted from [5])

to the energy stored. It should be noted that the loss tangent at zero electric field (approximately 0.1 for 25 Hz) is higher than the loss tangent reported by the supplier (i.e. 0.013). This is due to continued domain wall motion at zero electric field; if the electric field is held constant at zero, there is a transient response over approximately 1 s where ferroelectric domains reach their equilibrium configuration and the loss tangent decreases to 0.02, which is closer to the reported value from the manufacturer (see [5] for a more detailed discussion of this phenomenon).

Bending frequencies beyond those shown in Figs. 10.3 and 10.4 were also tested. The relative compliance (inverse of Young's modulus) and loss tangent versus frequency are shown in Fig. 10.5. In particular, Fig. 10.5a shows the compliance measured at two points during the cyclic electric field: one at zero electric field and one at the coercive field (where there is the maximum amount of softening). Both curves are relative to the compliance measured at 25 Hz and zero electric field. The compliance at zero electric field shows the typical resonance and anti-resonance behavior; the compliance (proportional to the amplitude of the specimen's deflection) increases at the first resonance frequency at approximately 120 Hz and then decreases at the first anti-resonance frequency at 250 Hz. The behavior continues for the higher harmonics. The measurement taken at the coercive field shows a similar behavior except that the resonance behavior tends to be shifted to lower frequencies. This is due to the softening of the Young modulus causing a resulting decrease in the specimen's (anti-)resonance frequencies. The loss tangent shown in Fig. 10.5b shows the well-known behavior of large increases in damping at resonance and anti-resonance frequencies. Furthermore, the loss tangent at the coercive field is higher than the loss tangent at zero electric field for a wide range of frequencies.

10.3.2 *Torsion Experiments*

Going beyond previous studies considering only bending [4] or uniaxial compression [15], BES allows for multi-axial testing of specimens in both bending (via vertical coils) and torsion (via horizontal coils). Torsion tests were performed to measure the relative shear modulus and loss tangent for the same mechanical frequencies and cyclic electric field in Figs. 10.3 and 10.4. The qualitative behavior of the relative shear modulus and loss tangent is the same as observed for bending. That is, there is a linear change in the modulus and loss tangent away from domain switching due to the electric field dependence of the elastic constants. In addition, during domain switching there is softening due to the anisotropy of elastic constants as well as softening due to relaxation during mechanical cycling. Similarly, an increase in loss tangent during domain switching arises due to domain wall motion and its associated dissipation. For brevity, the equivalent plots of Figs. 10.3 and 10.4 for torsion are not shown and instead the reader is referred to [5]. However, the equivalent plots for Fig. 10.5 in torsion are shown in Fig. 10.6. For the specimen tested, the resonance frequency in torsion was much higher than in bending, thus there is a large increase in compliance and loss tangent at approximately 1300 Hz. It can also be seen in Fig. 10.6b that the increase in loss tangent during domain switching in torsion occurs for a wide range of torsional frequencies (as was the case in bending).

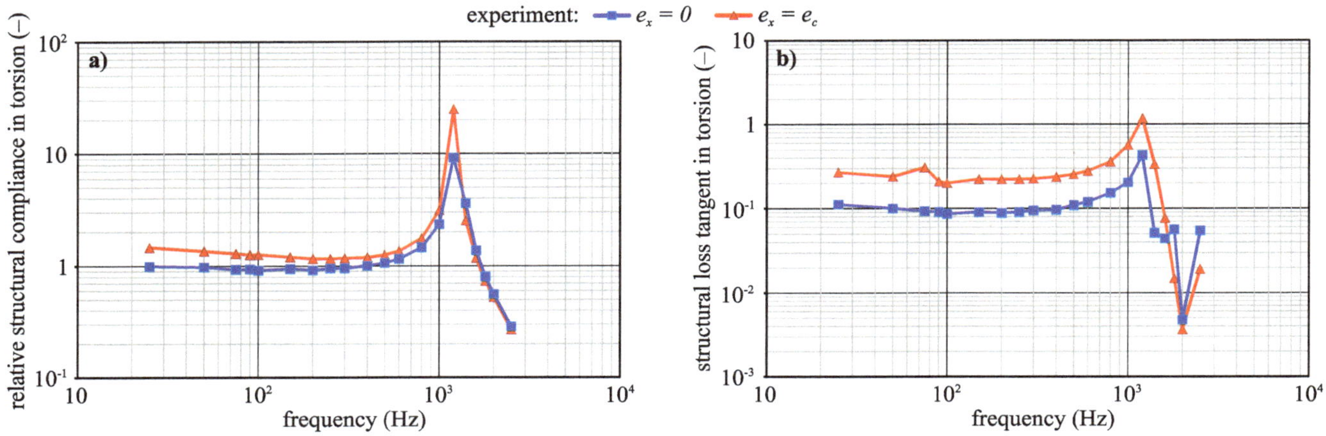

Fig. 10.6 (**a**) Relative compliance and (**b**) loss tangent versus torsional frequency measured at zero electric field ($e_x = 0$) and at the coercive field ($e_x = e_c$) (Adapted from [5])

10.4 Conclusion

The new method and apparatus of BES presented allows for the application of mechanical and torsional moments while simultaneously applying cyclic electric fields. The method was used for characterizing the evolution of the dynamic moduli and loss tangent of PZT over a wide range of frequencies beyond those studied previously. Softening of the Young and shear moduli as well as large increases in their associated loss tangents were observed near the coercive field when domain switching is occurring. Due to the wide range of bending and torsional frequencies accessible with BES, the dynamic stiffness and loss tangent were measured through the full resonance spectrum of the specimen. The increased loss tangent (both in bending and torsion) during domain switching occurred for a wide range of frequencies. This result reveals that domain switching is a mechanism in ferroelectrics that leads to broadband damping as opposed to exhibiting high damping only for narrow frequency ranges (as in previous techniques that only utilize the linear piezoelectric response). Furthermore, by significantly enhancing the damping capacity of stiff ferroelectric ceramics using electric fields, materials with tunable high stiffness and high damping were achieved.

Acknowledgements Financial support from United Technologies Research Center as well as from the Caltech Innovation Initiative (CI2) is gratefully acknowledged.

References

1. Hagood, N.W., von Flotow, A.: Damping of structural vibrations with piezoelectric materials and passive electrical networks. J. Sound Vib. **146**(2), 243–268 (1991)
2. Asare, T.A., Poquett, B.D., Schultz, J.P., Kampe, S.L.: Investigating the vibration damping behavior of barium titanate ($BaTiO_3$) ceramics for use as a high damping reinforcement in metal matrix composites. J. Mater. Sci. **47**(6), 2573–2582 (2012)
3. Arlt, G., Dederichs, H.: Complex elastic, dielectric and piezoelectric constants by domain wall damping in ferroelectric ceramics. Ferroelectrics **29**(1), 47–50 (1980)
4. Jiménez, B., Vicente, J.M.: Influence of mobile 90° domains on the complex elastic modulus of PZT ceramics. J. Phys. D Appl. Phys. **33**(12), 1525 (2000)
5. le Graverend, J.-B., Wojnar, C.S., Kochmann, D.M.: Broadband Electromechanical Spectroscopy: characterizing the dynamic mechanical response of viscoelastic materials under temperature and electric-field control in a vacuum environment. J. Mater. Sci. **50**, 3965–3984 (2015)
6. Wojnar, C.S., le Graverend, J.-B., Kochmann, D.M.: Broadband control of the viscoelasticity of ferroelectrics via domain switching. Appl. Phys. Lett. **105**(16), 162912 (2014)
7. Tamura, M., Ogasawara, K., Ono, N., Hagiware, S.: Piezoelectricity in uniaxially stretched poly(vinylidene fluoride). J. Appl. Phys. **45**(9), 3768–3771 (1974)
8. Lakes, R.S.: Viscoelastic measurement techniques. Rev. Sci. Instrum. **75**(4), 797–810 (2004)
9. Lakes, R.S.: Viscoelastic Solids. CRC, Boca Raton (1998)
10. Sawyer, C.B., Tower, C.H.: Rochelle salt as a dielectric. Phys. Rev. **35**, 269–273 (1930)
11. Zhou, D., Kamlah, M., Munz, D.: Rate dependence of soft PZT ceramics under electric field loading. Proc. SPIE, **4333**, 64–70 (2001)

12. Fu, R., Zhang, T.-Y.: Effects of an electric field on the fracture toughness of poled lead zirconate titanate ceramics. J. Am. Ceram. Soc. **83**(5), 1215–1218 (2000)
13. Burlage, S.R.: The dependence of elastic constants on polarization in a ferroelectric ceramic. IEEE Trans Sonics Ultrason. **12**(1), 5–8 (1965)
14. Wang, Q.-M., Zhang, T., Chen, Q., Du, X.-H.: Effect of DC bias field on the complex materials coefficients of piezoelectric resonators. Sensors Actuators A Phys. **109**, 149–155 (2003)
15. Chapyla, P.M., Carman, G.P.: Compression of piezoelectric ceramic at constant electric field: energy absorption through non-180° domain-wall motion. J. Appl. Phys. **92**(3), 1504–1510 (2002)

Chapter 11
Dynamics of Microscale Granular Crystals

Nicholas Boechler

Abstract Granular media support a rich array of acoustic phenomena that stem from their complex microstructure and highly nonlinear particulate interactions. While such complexity enables granular media's unique properties, it simultaneously makes predicting their dynamic response challenging. An approach that has yielded insights into the dynamics of macroscale granular media is to study ordered and reduced-dimensional granular systems. Such simplified granular systems are commonly referred to as "granular crystals." Granular media composed of micro- and nanoscale particles are predicted to be analogous in many ways to their macroscale counterparts, however, many phenomena that are negligible at macroscales, such as interparticle adhesion, become critical at sub-microscales. In this chapter, we review several recent studies of the contact-based dynamics of microscale granular crystals. This includes measurements of the interactions of guided and bulk acoustic waves with the contact resonances of self-assembled microsphere monolayers. These works have implications for the broader study of microscale granular materials, provide new ways to study microscale contact mechanics, and may result in a new class of materials for passive wave tailoring, which can be rapidly and inexpensively fabricated in large scales via self-assembly.

Keywords Granular media • Laser ultrasonics • Contact mechanics

Granular materials consist of assemblies of deformable particles in contact. From sand to powders, granular media are ubiquitous in industry and our natural environment [1]. The study of how dynamic disturbances, including acoustic and other mechanical waves, propagate through granular media is an active field of research with applications to areas such as non-destructive testing, wave propagation in geological materials, and powder processing [2]. Granular media are known to support a rich array of dynamic phenomena including highly nonlinear mechanical wave propagation [2] and jamming [3], which stem from their complex, often heterogeneous, structure and highly nonlinear particulate interactions. This combination of nonlinearity and structural heterogeneity has also historically made predicting and understanding the dynamic response of granular media challenging.

One approach that has provided an avenue for new understanding of granular media dynamics is to study ordered or reduced dimensional granular systems, which are commonly referred to as "granular crystals." Granular crystals, as shown in Fig. 11.1, are composed of arrays of uniform (or otherwise intentionally disordered) arrays of spherical particles [4, 5] that are typically modeled to interact according to the Hertzian contact model [6]. For dynamic disturbances with timescales that are significantly less than the intrinsic spheroidal resonances of isolated particles, the spheres can be thought to vibrate like rigid-bodies with local deformation around the point of contact that acts like a spring [2]. In addition to yielding insights regarding granular media dynamics, granular crystals can be thought of as a type of nonlinear phononic crystal or metamaterial, wherein the nonlinear contact interactions and discreteness-induced dispersion yield novel wave phenomena such as nonlinear intrinsic localized modes and non-reciprocal acoustic propagation [4]. While granular crystals have been a topic of major interest and yielded an array of novel phenomena, prior investigations have been largely restricted to systems composed of macroscale particles.

The dynamics of granular crystals containing particles of $\sim$1 μm diameter and below have fundamental physical differences from the dynamics of macroscale granular crystals. At macroscales, the contact between particles has been shown to be well described by Hetzian contact interactions. Shear interactions also play a role (particularly in statically compressed granular systems), and are typically modeled using the Hertz-Mindlin contact model [7]. The presence of shear interactions has been shown to result in the existence of unique phenomena such as rotational modes [8]. When considering the contact

N. Boechler (✉)
Department of Mechanical Engineering, University of Washington, Box 352600, Seattle, WA 98195, USA
e-mail: boechler@uw.edu

L. Starman et al. (eds.), *Micro and Nanomechanics*, Conference Proceedings of the Society for Experimental Mechanics Series,
DOI 10.1007/978-3-319-42228-2_11

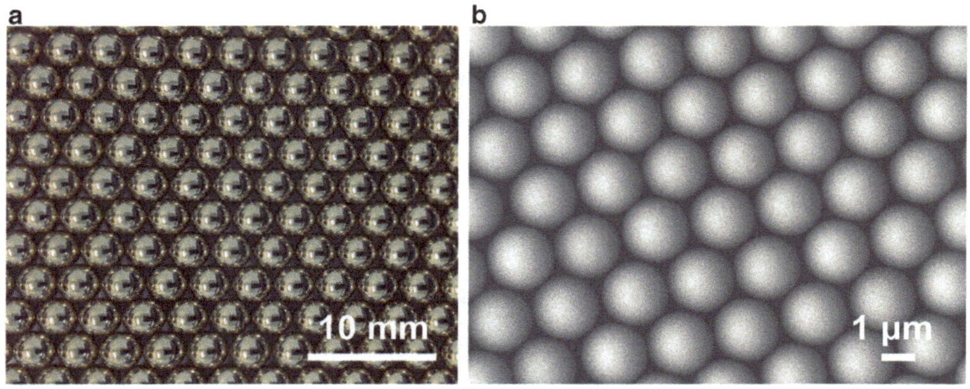

Fig. 11.1 2D (**a**) macro- and (**b**) microscale granular crystals

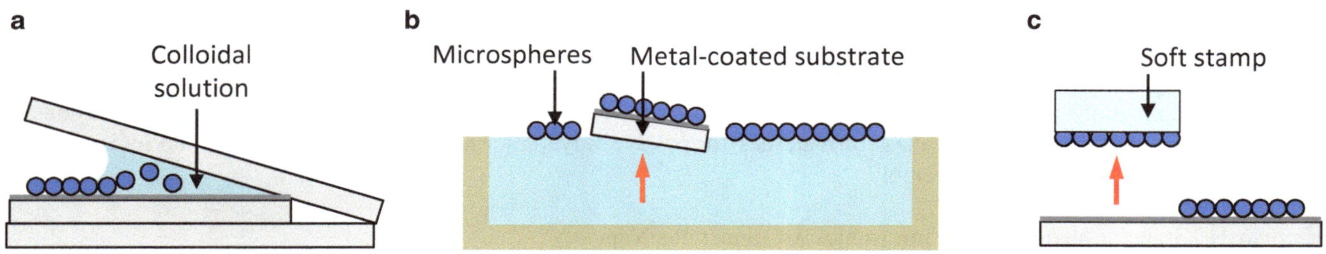

Fig. 11.2 (**a**) Wedge-cell and (**b**) Langmuir-Blodgett 2D microscale granular crystal self-assembly fabrication methods. (**c**) Microcontact printing method

between microscale spheres, adhesion, due predominantly to van der Waals forces, plays a major role and must be taken into account [9]. Because of the strong static compression due to adhesive forces, shear interactions and particle rotations are predicted to play an even greater role than in macroscale granular crystals. In addition, effects due to surface roughness [9] and plastic deformation may also play a greater role (the later due to the strong static compression from adhesive forces) [10]. In general, the study of microscale contact mechanics is still an open field of study, as previous studies regularly show large variation between experiments [11] and between experiments and theoretical predictions [9].

In addition to the new physical effects that should be taken into account when considering the dynamics of microscale granular crystals, the small particle size also yields new experimental challenges. This includes higher, megahertz to gigahertz, contact resonance frequencies, smaller characteristic displacements (for instance, the static overlap between microspheres is typically less than a few nanometers), and difficulties in the manufacture of ordered microscale granular structures. In light of these challenges, convective self-assembly fabrication and laser-ultrasonic characterization techniques have proven effective in the, respective, manufacture and characterization of microscale granular crystals.

Drawing from the field of colloidal assembly [12], several convective self-assembly techniques have been utilized in recent studies of microscale granular crystal dynamics. A wedge-shaped-cell convective self-assembly technique [13], as shown in Fig. 11.2a, was utilized to assembly monolayers of 1–2 μm diameter silica particles on metal-coated glass microscope slides [14, 15]. A modified Langmuir–Blodgett technique [16], wherein polystyrene (PS) particles are assembled into a monolayer structure at an air/water interface and then transferred to another substrate, as shown in Fig. 11.2b, has also been utilize to construct two-dimensional (2D) microscale granular crystals on ultra-thin silicon membranes [17] and metal-coated glass slides [18, 19]. An advantage of this technique is that the substrate and particle size can be easily modified, with little modification to the assembly conditions. Previous studies have utilized microcontact printing [20] to create flat interfaces [15] and strips [19] of 2D granular crystals adhered to substrates, as shown in Fig. 11.2c.

Several laser ultrasonic techniques have been utilized to characterize the dynamic response of microscale granular crystals, including laser-induced transient grating spectroscopy [21] and other "pump-probe" techniques, each of which utilize a short pulse laser as a "pump," and a continuous wave (CW) laser as a "probe" [14, 15, 17–19]. The absorption of pump light by a metal transducer film on the substrate causes a rapid thermal expansion that launches acoustic waves in the substrates and the microscale granular crystals. Intensity changes of the probe light induced by acoustic waves in the samples are measured with fast photodiodes, digitized with high-speed oscilloscopes, and averaged over multiple acquisitions. In the laser-induced transient grating spectroscopy technique [21], as shown in Fig. 11.3a, the pump beam is split in two using

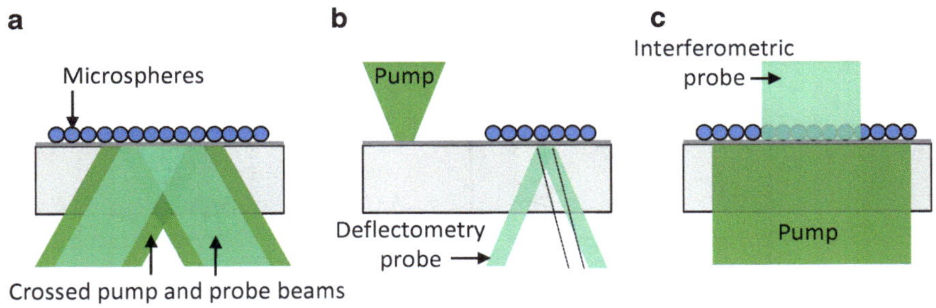

Fig. 11.3 Schematics of laser ultrasonic characterization methods utilized to study the dynamics of 2D microscale granular crystals. (**a**) Laser-induced transient grating spectroscopy. (**b**) Line-source pump excitation and photo-deflectometry probe. (**c**) Large pump spot excitation and interferometric probe

a phase mask and then crossed at the sample surface to form an interference pattern. This interference pattern launches narrowband (in spatial frequency) counter-propagating surface acoustic waves (SAWs) with wavelength equal to the period of the interference pattern. A modulated CW laser is split through the same phase mask and focused on the sample surface, where surface ripples and refractive index variations induced by the SAWs in the sample that form a "transient grating" diffract the one of the beams. The diffracted beam is overlapped with the second half of the probe beam, which is reflected off of the sample. Time evolution of the SAWs in the substrate result in intensity changes in the overlapped probe beam. This technique has been utilized to measure the dispersion and identify hybridization of SAWs and Lamb waves with the out-of-plane microsphere contact resonance of a 2D microscale granular crystal adhered to a thick [14, 18] and thin substrates [17], respectively. This technique has also been utilized to study the resonant attenuation of SAWs propagating across a strip of 2D microscale granular crystal adhered to a substrate [19]. Other laser-ultrasonic experimental configurations utilized to study the dynamics of microscale granular crystals are shown in Fig. 11.3b, c. Figure 11.3b shows a configuration that was utilized to study the resonant attenuation of SAWs in a 2D microscale granular crystal adhered to a substrate [15]. In this configuration, the pump beam is focused into a line source, which launches broadband counter propagating SAW pulses. One of the pulses crosses an interface at the edge of the 2D microscale granular crystal. By taking measurements with varied probe positions on the sample, the resonant attenuation and dispersion of SAWs, due to presence of the microscale granular crystal, can be measured. Figure 11.3c shows a configuration where a large spot (compared to the microsphere dimension) pump beam is focused on the transducer layer. Due to symmetry constraints, this configuration preferentially excites out-of-plane modes of the microscale granular crystal. The resulting microsphere vibrations are detected by a grating interferometer [22], which is similarly most sensitive to out-of-plane motion of the microspheres. This provides an avenue to differentiate between different modes, with in- or out-of-plane motion, of the microscale granular crystal [15]. In any of the experimental techniques shown in Fig. 11.3, measurements can be made with either the pump or the probe incident from the side of the microscale granular crystal or the substrate. Configuring the measurement such that the probe is incident on the side of the microscale granular crystal versus the substrate side provides greater sensitivity to the motion of the spheres and substrate, respectively [14].

In the works of [14, 17, 18], the laser induced transient grating spectroscopy technique was utilized to measure the dispersion of guided acoustic waves through a system composed of 2D microscale granular crystals adhered to substrates. In each of these studies, the presence of a single contact resonance was detected via hybridization with the guided acoustic waves. These measurements matched well with the predictions of a simple model with non-interacting oscillators having a single, out-of-plane degree of freedom, as shown in the schematic of Fig. 11.4a. While the model well captured the observed dynamics, effects related to microsphere motion in the in-plane direction, particle rotations, and interparticle coupling due to adhesion between the microspheres were not observed. A later theoretical work developed a more advanced model that took these effects into account, using the model shown in Fig. 11.4b [23]. This work revealed several major findings. Due to the possibility of in-plane (transverse) motion of the microspheres and particle rotations, a 2D microscale granular crystal adhered to a substrate should have two additional contact resonances (three in total, counting the previously observed out-of-plane resonance) that hybridize with SAWs in the substrate and have a combination of translational and rotational motion. Without the presence of particle rotations, the model predicts only the out-of-plane contact resonance and a second, purely translational resonance contact, with in-plane motion.

Following the works of [14, 17, 18], an experiment was conducted that utilized the experimental configuration shown in Fig. 11.3b and studied the resonant attenuation of SAWs traversing an interface between regions with and without an adhered 2D microscale granular crystal [15]. The Fourier spectra of the measurements presented in [15] revealed the presence

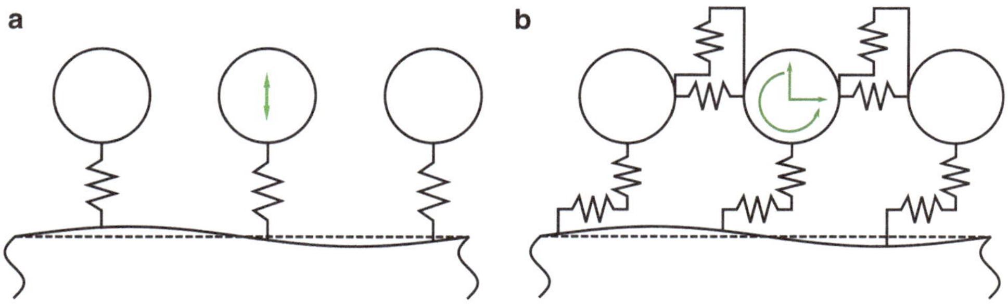

Fig. 11.4 Models used to describe the dynamics of 2D microscale granular crystals adhered to deformable substrates. (**a**) Model with non-interacting spheres and a single degree of freedom. (**b**) Model with interacting spheres and three degrees of freedom, including particle rotations. Panels (**a**) and (**b**) are adapted from Ref. [23]

of three attenuation zones, which were attributed to the three contact resonances predicted in [23]. The top and bottom attenuation zones were of weaker strength, which suggested that they are more weakly coupled resonances and is consistent with their lack of observation in [14, 17, 18]. We note that in the supplementary of [14], a drop in the amplitude of the peak corresponding to the Rayleigh-like mode in one measured spectrum near the predicted frequency of the lower resonance may have also similarly indicated the presence of this resonance.

To confirm the nature of each of the three observed resonances, the interparticle contact stiffness was increased by evaporating a thin layer of 20 and 40 nm of aluminum onto the top of the 2D microscale granular crystal [15]. Because the aluminum was deposited from the top, the microsphere-substrate contact was in the "shadow" of the deposition, and remained unaffected. Upon deposition of the aluminum, the resonances with the highest and lowest frequencies (that were predicted to correspond to the transverse-rotational resonances) shifted to higher frequencies. The middle resonance that was attributed to the out-of-plane contact resonance downshifted slightly, which is consistent with the shift predicted due to mass loading from the aluminum. In each case, the measured frequency of the middle resonance, and the model of [23], was used to find the particle-substrate normal contact stiffness. The particle-substrate shear contact stiffness was then found using the Hertz-Mindlin contact model from the measured particle-substrate normal contact stiffness, and the interparticle shear stiffness using a least squares fit to the measured transverse-rotational resonances. Qualitative agreement between the measurements and the model confirmed the nature of each of the predicted and observed contact resonances. Differences in the transverse-rotational frequencies were attributed to deviations from the model of [23] due to the presence of the aluminum, including affects such as asymmetry and the formation of necks that may result in increased bending stiffness. As a secondary confirmation, an experiment using the configuration shown in Fig. 11.3c was conducted [15]. The measured Fourier spectra showed the presence of a single peak a frequency that matched the previously identified out-of-plane contact resonance.

One of the important ramifications of these types of dynamic experiments are that they provide new information regarding the contact stiffness of adhered microparticles around the equilibrium position, using a non-destructive approach. One of the most common prior methods to investigate microsphere contact mechanics and adhesion is to use an atomic force microscope (AFM) to study the force at which microspheres detach from the surface they are adhered to [9]. While measurements of the pull-off force have been widely used, effects due to the contact of the AFM cantilever must be taken into account and the measurement provides little information of the contact stiffness near equilibrium. In the previously described experiment that studied resonant attenuation of SAWs [15], and other experiments on, 2D microscale granular crystals [14, 15, 17–19], the particle-substrate contact stiffness was found to be over twice as large as predicted by adhesive elastic contact models [9]. This discrepancy was attributed to several factors. Uncertainty in the work of adhesion may cause underpredicted contact stiffnesses. For instance, particularly in the case of dielectric-metal adhesion, significantly larger than predicted works of adhesions were measured for reactive metals, such as aluminum [24]. Plastic deformation at the contact may also contribute to this upshift, as a larger contact area can result in larger stiffness than would be predicted by elastic contact models [10]. In addition to the higher than predicted particle-substrate stiffnesses, the ratio of interparticle to particle-substrate contact stiffness was found to be over five times smaller than predicted in Ref. [15]. This discrepancy was explained by observations from scanning electron microscope images, which showed that the particles are typically not in contact with all six of their neighbors [15], which would result in a lower than predicted average interparticle contact stiffness.

The studies reviewed in this chapter provide the foundation for the emerging field of microscale granular crystal dynamics. In particular, the study of Ref. [15] provided the first observation of the interparticle-mediated contact-based dynamics of a two-dimensional microscale granular crystal, including the observation of modes that involving both displacement and rotation of the particles. As with their macroscale counterparts, microscale granular crystals may find use as a type of designer material in applications that leverage their potential for passive stress and acoustic wave tailoring [4]. However, in contrast to their macroscale counterparts, microscale granular crystals may be manufactured quickly and in large quantities by utilizing self-assembly fabrication techniques. Furthermore, because of their small scales, microscale granular crystals may find use in applications where size is a critical issue, such as SAW signal processing and sensing devices, and may yield novel acousto-optic or acousto-plasmonic properties. Finally, these studies of microscale granular crystal dynamics [14, 15, 17–19] and the findings therein, such as the effects of self-assembled interparticle contact network formation on effective interparticle stiffness [15], provide improved understanding of wave propagation in microgranular media with applications to shock and impact mitigation, energetic material dynamics, seismic exploration, and powder processing [2]. These investigations form the first steps towards future studies of both higher- and lower-dimensional microscale granular crystals, including studies of microscale granular crystal dynamics at short wavelengths and highly nonlinear regimes, and studies of microscale granular crystals that have greater degrees of structural complexity and hierarchy.

Acknowledgments N.B. acknowledges the collaborators on the projects detailed in this work: M. Hiraiwa, A. Khanolkar, S. P. Wallen, M. Abi Ghanem, J. Jenks, J. K. Eliason, A. Vega-Flick, A. A. Maznev, K. A. Nelson, A. Kumar, T. Gan, N. Fang, and N. Vogel. This work was supported by the US National Science Foundation (grant no. CMMI-1333858) and the US Army Research Office (grant no. W911NF-15-1- 0030).

References

1. Duran, J.: Sands, Powders, and Grains: An Introduction to the Physics of Granular Materials. Springer-Verlag, New York (2000)
2. Nesterenko, V.F.: Dyamics of Heterogeneous Materials. Springer, New York (2001)
3. Bi, D., Zhang, J., Chakraborty, B., Behringer, R.P.: Jamming by shear. Nature **480**, 355 (2011)
4. Theocharis, G., Boechler, N., Daraio, C.: Nonlinear periodic phononic structures and granular crystals. In: Acoustic Metamaterials and Phononic Crystals, pp. 217–251. Springer, Heidelberg (2013)
5. Porter, M.A., Kevrekidis, P.G., Daraio, C.: Granular crystals: nonlinear dynamics meets material engineering. Phys. Today **68**, 44 (2015)
6. Hertz, H.: On the contact of elastic solids. Reine Angew. Math. **92**(156) (1882)
7. Mindlin, R.D.: Compliance of elastic bodies in contact. J. Appl. Mech. **16**, 249 (1949)
8. Merkel, A., Tournat, V., Gusev, V.: Experimental evidence of rotational elastic waves in granular phononic crystals. Phys. Rev. Lett. **107**, 225502 (2011)
9. Israelachvili, J.N.: Intermolecular and Surface Forces. Academic, Waltham (2011)
10. Maugis, D., Pollock, H.M.: Surface forces, deformation and adherence at metal microcontacts. Acta Metall. **32**, 1323 (1984)
11. Vittorias, E., Kappl, M., Butt, H.-J., Johannsmann, D.: Studying mechanical microcontacts of fine particles with the quarts crystal microbalance. Powder Technol. **203**, 489 (2010)
12. Vogel, N., Retsch, M., Fustin, C.-A., del Camp, A., Jonas, U.: Advances in colloidal assembly: the design of structure and hierarchy in two and three dimensions. Chem. Rev. **115**, 6265 (2015)
13. Sun, J., Tang, C.J., Zhan, P., Han, Z.I., Cao, Z.S., Wang, Z.L.: Fabrication of centimeter-sized single-domain two-dimensional colloidal crystals in a wedge-shaped cell under capillary forces. Langmuir **26**, 7859 (2010)
14. Boechler, N., Eliason, J., Kumar, A., Maznev, A., Nelson, K., Fang, N.: Interaction of a contact resonance of microspheres with surface acoustic waves. Phys. Rev. Lett. **111**, 036103 (2013)
15. Hiraiwa, M., Abi Ghanem, M., Wallen, S.P., Khanolkar, A., Maznev, A.A., Boechler, N.: Complex contact-based dynamics of microsphere monolayers revealed by resonant attenuation of surface acoustic waves. Phys. Rev. Lett. **116**, 198001 (2016)
16. Vogel, N., Goerres, S., Landfester, K., Weiss, C.K.: A convenient method to produce close- and non-close- packed monolayers using direct assembly at the air-water interface and subsequent plasma-induced size reduction. Macromol. Chem. Phys. **212**, 1719 (2011)
17. Khanolkar, A., Wallen, S.P., Abi Ghanem, M., Jenks, J., Vogel, N., Boechler, N.: A self-assembled metamaterial for Lamb waves. Appl. Phys. Lett. **107**, 071903 (2015)
18. Vega-Flick, A., Eliason, J.K., Maznev, A.A., Khanolkar, A., Abi Ghanem, M., Boechler, N., Alvarado-Gil, J.J., Nelson, K.A.: Laser-induced transient grating setup with continuously tunable period. Rev. Sci. Instrum. **86**, 123101 (2015)
19. Eliason, J.K., Vega-Flick, A., Hiraiwa, M., Khanolkar, A., Gan, T., Boechler, N., Fang, N., Nelson, K.A., Maznev, A.A.: Resonant attenuation of surface acoustic waves by a microgranular metamaterial. Appl. Phys. Lett. **108**, 061907 (2016)
20. Lee, B.H., Shin, H., Sung, M.M.: Patterning a two- dimensional colloidal crystal by water-mediated particle transfer printing. Chem. Mater. **19**, 5553 (2007)
21. Rogers, J.A., Maznev, A.A., Banet, M.J., Nelson, K.A.: Optical generation and characterization of acoustic waves in thin films: fundamentals and applications. Annu. Rev. Mater. Sci. **30**, 117 (2000)
22. Glorieux, C., Beers, J.D., Bentefour, E.H., Van de Rostyne, K., Nelson, K.A.: Phase mask based interferometer: operation principle, performance, and application to thermoelastic phenomena. Rev. Sci. Instrum. **75**, 2906 (2004)
23. Wallen, S.P., Maznev, A.A., Boechler, N.: Dynamics of a monolayer of spheres on an elastic substrate. Phys. Rev. B **92**, 174303 (2015)
24. Lipkin, D.M., Israelachvili, J.N., Clarke, D.R.: Estimating the metal-ceramic van der Waals adhesion energy. Philos. Mag. A **76**, 715 (1997)

FSC
www.fsc.org
MIX
Papier | Fördert
gute Waldnutzung
FSC® C083411

Kolibri 360 GmbH
Ferdinand-Jühlke-Straße 7
99095 Erfurt, Deutschland
produktsicherheit@kolibri360.de